INSTITUTE OF PSYCHIATRY

Maudsley Monographs

MAUDSLEY MONOGRAPHS

HENRY MAUDSLEY, from whom the series of monographs takes its name, was the founder of the Maudsley Hospital and the most prominent English psychiatrist of his generation. Maudsley Hospital was united with the Bethlem Royal Hospital in 1948, and its medical school, renamed the Institute of Psychiatry at the same time, became a constituent part of the British Postgraduate Medical Federation. It is entrusted by the University of London with the duty to advance psychiatry by teaching and research.

The monograph series reports work carried out in the Institute and in the associated Hospital. Some of the monographs are directly concerned with clinical problems; others, less obviously relevant, are in scientific fields that are cultivated for the furtherance of psychiatry.

Editors

INSTITUTE OF PSYCHIATRY

Maudsley Monographs

Number Thirty-Three

The Epidemiology of Childhood Hyperactivity

By

ERIC TAYLOR, FRCP, FRCPsych
*Reader in Developmental Neuropsychiatry, Institute of Psychiatry,
Member of Medical Research Council Scientific Staff,
Honorary Consultant in Child and Adolescent Psychiatry, Bethlem Royal and
Maudsley Hospitals, and King's College Hospital, London.*

SEIJA SANDBERG, Lic Med (Finland), MD, FRCPsych
*Senior Lecturer in Child and Adolescent Psychiatry, University of Glasgow,
Honorary Consultant, Royal Hospital for Sick Children, Glasgow.*

GEOFFREY THORLEY, BA, MPhil, PhD, AFBPS
Consultant Clinical Psychologist, Leicester Health Authority.

SUSAN GILES, BA
Research Worker, Institute of Psychiatry, London.

OXFORD UNIVERSITY PRESS

1991

Oxford University Press, Walton Street, Oxford OX2 6DP
Oxford New York Toronto
Delhi Bombay Calcutta Madras Karachi
Petaling Jaya Singapore Hong Kong Tokyo
Nairobi Dar es Salaam Cape Town
Melbourne Auckland
and associated companies in
Berlin Ibadan

Oxford is a trade mark of Oxford University Press

Published in the United States
By Oxford University Press, New York

A catalogue record for this book is available from the British Library

Library of Congress Cataloging in Publication Data
The Epidemiology of childhood hyperactivity / by Eric Taylor . . . [et al.].
pp. cm.—(Maudsley monographs; no. 33)
Includes bibliographical references and index.
1. Hyperactive child syndrome—Epidemiology. 2. Conduct disorders
in children—Epidemiology. I. Taylor, Eric A. II. Series.
[DNLM: 1. Attention Deficit Disorder with Hyperactivity—epidemiology.
W1 MA997 no. 33 / WS 350.6 E635]
RJ506.H9E75 1992 614.5'98589—dc20 91–3889
ISBN 0-19-262172-6

Set by
Dobbie Typesetting Limited, Tavistock, Devon
Printed in Great Britain by
Bookcraft Ltd., Midsomer Norton, Avon

Acknowledgements

The research reported here was funded by the Medical Research Council and we are appreciative to them and to the MacArthur Foundation for support. We have been indebted throughout to Professor Michael Rutter for his wise guidance and encouragement. Drs Mary Anne Griffiths and Leslie Davidson contributed greatly to specific aspects of the study, as noted in the text. A major part of the production of this book has fallen to Catherine Buckley and we want to record the extent of the contribution, the efficiency of it and our gratitude. Our deepest debt of gratitude, as will be evident from the book, goes to the children, their families and their schools. They have helped us unstintedly and we hope that their successors will benefit from what they have taught us.

Contents

Tables

Figures

Outline and summary

The idea behind this study was that the large, heterogeneous grouping of antisocial conduct disorders in children contains a subgroup of children with hyperkinesis, caused by dysfunction of the brain rather than by psychosocial problems.

Most previous research work has found it hard to address this issue sharply, largely because of problems in the definition of cases. Most definitions of hyperactivity are very inclusive, allow a large proportion of children to receive the diagnosis, include antisocial problems in the definition—and generate few useful predictions. Most studies compare clinic cases of hyperactivity with children from the general population, so confounding the causes of hyperactivity with the reasons for referral. These studies will be referred to in the context of methodological issues in the following chapters.

In previous work with clinically referred cases we have developed methods for the assessment of hyperactivity and other behaviour problems and for measuring disturbance of attention and neurological development. Statistical taxonomic studies using these methods indicated that separate dimensions of hyperactivity and conduct disorders should be distinguished; that a group of children with pervasive hyperactivity and inattention should be recognized; and that the hyperkinetic group is further characterized by neurodevelopmental delays and a marked response to stimulant medication.

The present study applied the methods and concepts from the clinical work in an epidemiological setting. It set out to identify and compare groups of hyperactive children and conduct-disordered children who showed the problems in pure form; and also children with both problems mixed together and a normal control group. We hypothesized that the associations of the two types of disorder would be different, hyperactivity being linked to impaired attention and conduct disorder to psychosocial problems. We also aimed to examine the prevalence and associations of the separate group of hyperkinetic disorder.

The investigation used a multiple-stage design, in which a community sample in East London was screened with rating scales by teachers and parents; comparison groups were selected on the basis of their rating scale scores for intensive study; and diagnostic groups were defined on the basis of the intensive measures and compared on a range of cognitive, neurological, and psychosocial measures.

Chapter 2 reports the first stage of the inquiry, the questionnaire data on the entire sample; and addresses the issue of how far descriptive ratings support the separation of distinct subtypes of disruptive behaviour.

There were 3215 boys on school rolls whose birthdates were within the range of the study (plus another 8 who were at units for children with severe learning delays and were therefore excluded from this stage of the survey). Teacher rating-scales were obtained for about 99 per cent and parent rating scales for about 80 per cent. Non-respondent parents had slightly more hyperactive children, but the sample was reasonably representative. Disruptive behaviour was common, and components of hyperactivity and conduct disorder could be separated. Many children met criteria for both types of disorder; these mixed cases had a more severe manifestation of both problems. Discriminant function analysis classified most of the mixed cases with the hyperactive rather than the conduct disordered group.

There was a tendency for the hyperactive children to show more symptoms of emotional disorder than other children. This, like the greater severity of the mixed cases, is a confounding factor that accounts for some of the contradictory results in the literature. Children were therefore excluded from subsequent stages of the study if they scored above a cut-off point on the scales of emotional symptomatology, and when this was done emotional disorder was no longer a confounding factor. Excluding boys with affective symptoms meant that the study did not examine their course. Accordingly, the relationship between conduct disorder and emotional disturbance is reported in Chapter 3. There was an overlap between conduct and emotional disorder, no bigger than that expected by chance. The mixed cases were classified, by discriminant function analysis, with the cases of conduct disorder. However, the mixed cases also showed a different profile of symptoms of emotional disorder from that of either pure group. High rates of depressive symptoms were characteristic. It seemed likely that depressive symptomatology represented another route into conduct disorder, separate from hyperactivity.

Chapter 4 then goes on to describe the next stage of the research. Subgroups of boys with different types of disruptive behaviour were selected for detailed study. It was possible to set up contrasting groups with pure hyperactivity, pure attention deficit, pure conduct disorder ('pure' in the sense that they were free of other types of disorder), mixed hyperactivity and conduct disorder, and a control group that did not meet criteria for any of these types of disorder. The groups were valid in that the detailed measures taken at the second stage confirmed that they differed from one another in the expected ways.

Hyperactive boys performed less well than the non-hyperactive conduct-disordered boys on tests of attention, and were more active and inattentive during testing. However, there were few other differences from normal controls. A group of boys with a mixture of hyperactivity and conduct disorder showed a more severe impairment of concentration and were delayed in learning to read. The biggest cognitive impairment was found in a group of non-hyperactive boys with attention deficit: they also showed lower verbal and performance IQ and poorer language development than controls. The cause did not seem to be connected with early injury to the brain or delays in motor development.

A similar story emerged from the systematic study of the developmental and health records that had previously been completed for nearly all the boys by the routine surveillance programme of the borough's Department of Community Health. The results are reported in Chapter 6. There were differences between groups in their language during their earlier development; and the problems were most clearly present in the non-hyperactive children who showed inattentive behaviour problems according to teachers' ratings. The boys who were later identified as hyperactive were more likely to have been recognized as showing problems of behaviour in early childhood. Chapter 6 also describes some predominantly negative findings about the role of allergies and accidents in the development of hyperactive children.

The comparisons of the same groups on standardized measures of family life and relationships and social background are reported in Chapter 5. The mixed HA-CD group was characterized by adverse factors in family relationships, and especially by high levels of negative expressed emotion towards the child by parents and low levels of skills in coping with behaviour problems. Depressive symptoms in mothers (though not full depressive syndromes) were also more common in this group. The findings were interpreted as a reaction to high intensity of behavioural disturbance, but may also be important in determining the later social adjustment of hyperactive children.

These results applied to the types of behavioural deviance identified by the questionnaires, not to individually diagnosed cases. They confirmed that hyperactivity was a useful construct for the description of disruptive children, and that it was particularly linked to cognitive changes. However, the associations were not strong, and certainly not sufficient to suggest that all boys with the behaviours of hyperactivity rated by adults should be seen as neuropsychiatrically impaired. Hyperactivity was continuously distributed in the population, as might be expected if indeed it results from multiple causes each of small effect. A prevalence rate would have reflected only the arbitrary decision of where to place the cut-off.

Accordingly, the detailed behavioural measures of the second stage were used to define a group of boys each of whom was characterized as showing hyperkinetic disorder. Operational research criteria were based upon the previous clinical studies in which the same measures had been employed. The level of definition was the same as that which had characterized a valid subgroup of clinically referred children as hyperkinetic. Chapter 7 reports the comparison of this group with two control groups: boys who were free of disruptive behaviour problems, and boys who were conduct disordered but not hyperactive. The prevalence of hyperkinetic disorder was approximately 1.7 per cent in 7- to 8-year-old boys (excluding those with known, severe learning disability). This was about one-tenth the rate of the DSM-III diagnosis of ADDH (Attention Deficit Disorder with Hyperactivity). The hyperkinetic disorder was separable from non-hyperactive conduct disorder by its early onset and frequent association with cognitive impairment, motor clumsiness, language delay and perinatal risk. Like those with conduct disorder, affected children were often living in families characterized by high levels of critical expressed emotion, low levels of coping, inconsistency between parents, and depressed mothers. The non-hyperactive but conduct-disordered children were specifically characterized by frequent marital discord between their parents and high rates of conduct disorder in first-degree relatives.

The restricted diagnosis of hyperkinetic disorder seems to be a useful category, similar to that seen in children with neurological disabilities. Its definition in this study is close to that proposed in the draft for the 10th revision of the International Classification of Diseases, and supports the proposal. Although restricted, and much less common than the frequent conditions of conduct disorder and ADDH, its frequency is still much greater than that diagnosed in current UK practice. Longitudinal studies and treatment trials will clarify the need for services for affected children. The wider category of children with hyperactive behaviour may still be at risk for persisting disorders of behaviour; the high rate of negative expressed emotion by parents argues that psychosocial interventions should be developed and assessed.

1 Introduction

DISRUPTIVE BEHAVIOURS AND THEIR CLASSIFICATION

Children are usually referred for psychiatric help because they are distressing others, rather than solely because of their own distress. The commonest single reason is that their behaviour persistently offends against accepted standards: they are violent to others, or destructive of property, or break the rules of society.

To be disruptive is not to be ill. One may choose to break rules, the explanation belonging to the moral or political domain rather than to the medical. For some children, however, rule-breaking is part of personal maladjustment. They show evidence of problems in personal development: perhaps persistent misery, or a lack of close friends, or an inability to learn. All advanced societies try to help such children to develop more happily. All recent psychiatric classification systems include a category of persistently rule-breaking behaviour as a type of deviance.

Several theories try to account for the development of these disorders of behaviour. One such approach is to consider them as evidence of inability to learn socialized activity—a deficit theory which can be formulated in terms of the biological functioning of the brain. Another approach emphasizes the failure of 'society' (usually personified as family members) to transmit the usual rules; whether by a lack of socialization or by a deviant subculture that transmits different rules. A third approach considers the motivation and preferences of the individual, determined by the interaction between constitutional traits and experiences of life. These approaches are not mutually exclusive. None of them has succeeded as a complete account, or driven the others from the field. Most reviewers consider that aetiology is complex, and that all three approaches contain some truth. In individual cases one process or another may be dominant, but in broadly defined groups there will be evidence of all.

The next step in studying the causes of disruptive disorders is therefore a classification of the disorders into subtypes. Of course, it is conceivable that no subtypes exist, and even that all cases are caused by the interaction of all factors. If, however, subtypes can be validly discriminated, then it is likely that their causes and associations will appear much more clearly than when all disorders are muddled together. For example, epidemiological studies have used parent and teacher ratings of children to distinguish conduct disorder from emotional disorder (Rutter, 1989*a*). When this was done, the strong associations of conduct disorder with other variables (such as reading disability and adverse family life) were revealed: they would not have appeared if all disturbed children had been studied as a single group.

The goal of a finer classification system for the conduct disorders has been pursued for many years. One of the earliest and simplest schemes is simply to specify the type of offence committed. Kanner's classic text, *Child Psychiatry* (1957), adopted this method, dealing successively with disobedience, lying, stealing, firesetting, destructiveness and cruelty, truancy from school, and running away. The disadvantages are considerable. From a scientific point of view, the types of offence have predicted little else of significance about the offenders. From a practical viewpoint, different types often co-exist and the number of antisocial behaviours shown by a child is a better guide to outcome than their type (Robins, 1978). More fundamentally, there is an objection to the idea that the particular rule transgressed is an important quality of the person. Before one can conclude that a disorder is present, one must be able to see that a specific aspect of personal functioning is impaired, rather than simply a reaction of the whole person.

A deep classification would embody an understanding of the causes of disorder. Scott (1963), for example, suggested a breakdown into groups, corresponding to the theories of aetiology mentioned earlier in this chapter. There would then be categories of children whose disorder was a failure to learn social rules, or a deviant socialization, or a maladaptive rule-breaking. Unfortunately, there is no clear means of assigning cases to categories, so the suggestion is not a scheme for classification, but only an assertion that classification is needed. Furthermore, it is not, on the face of it, probable that the categories would represent any natural ordering of cases; for it is unlikely that these aetiological factors operate independently.

Empirical studies have therefore been carried out by many investigators in order to generate subtypes of childhood disorders. Some of them are reviewed in later chapters in relation to particular questions. The general issues involved have been reviewed by Rutter and Garmezy (1983), Rutter and co-workers (1988), and Achenbach (1988). At this stage, only a few points need to be taken from the findings and commentaries. There are some promising indications that useful clinical distinctions can be made, and that the separation allows clearer understanding and more accurate prognosis. Socialized disorders have been separated in this way from unsocialized, aggressive from non-aggressive, delinquent from non-delinquent, oppositional from antisocial. There is, however, rather little agreement over the details of disorders to be recognized. One proposal is that a coherent subgroup of children is characterized by the presence of activity and attention deficit (Taylor, 1988).

CHILDHOOD HYPERACTIVITY

Hyperactivity is an enduring trait of inattentive, restless, uncontrolled behaviour. It can be considered as a *dimension* of behaviour, on which individuals can score higher or lower; but the distribution of individuals on the dimension has been little studied. DSM–III recognized a *category* of Attention Deficit Disorder, characterized by impulsivity and inattention, that could be subdivided into those with and those without the symptoms of motor 'hyperactivity' (ADDH and ADD) (American Psychiatric Association, 1980). DSM–III-R included motor overactivity as a requirement for the diagnosis, renamed it Attention Deficit-Hyperactivity Disorder (ADHD) and dropped the category of attention deficit without hyperactivity (American Psychiatric Association, 1987). ICD-9 described a condition of hyperkinetic syndrome, defined by 'extreme' overactivity and inability to attend, in the absence of other disorders (World Health Organization, 1978). ICD-10 proposed a grouping of hyperkinetic disorder, whose character-istics were severe and pervasive (cross-situational) inattentiveness and restlessness and the absence of pervasive developmental and affective disorders (World Health Organization, 1988). These alternative definitions are confusingly different, but share an assumption that psychiatry must recognize, treat, and study a persistent condition of early onset in which severe hyperactivity is prominent.

Disorders characterized by hyperactive behaviour are important because they are common and can be disabling. Their validity is supported by the consistent appearance of hyperactivity as a component in factor analytic studies of rating scales (see Chapter 2). Hyperactive disorders are of particular interest because of the idea that they form a biologically determined group. This idea has persisted ever since Still's (1902) descriptions of 'defects of moral control'. It has never been confirmed or disproved. Reviewers have generally concluded that no biological cause has been found except in a very few instances of structural damage to the brain; even the discriminative validity of the condition from the other disorders of disruptive conduct is doubtful

(Shaffer, 1980, Ferguson & Rapoport, 1983, Werry *et al.*, 1987, Rutter, 1989*b*, Taylor, 1988). Nevertheless, the conclusion—of a lack of evidence for biological causation—has always been qualified by the inadequacy of the evidence and the unsatisfactory nature of the diagnostic definitions.

The nature of the supposed deficit in hyperactivity is still unknown. Correspondingly, the definition of hyperactivity and the frequency with which it is applied vary between clinicians. Prendergast and co-workers (1988) showed that US diagnosticians used the diagnosis of hyperkinetic syndrome much more widely than those from the UK, even when all were rating the same cases. The diagnosis is used more widely by those who believe that biological deficits often contribute to disruptive behaviour and those who use pharmacological treatments. The result is that workers from different theoretical backgrounds confuse one another when they try to communicate.

Many UK clinicians see no reason to use the concept of hyper-activity. They prefer not to use the stimulant drugs anyway, so they have no pressing therapeutic reason for making the diagnosis; and they point to the lack of any demonstrated aetiological or prognostic homogeneity in children with ADHD. They do of course see children who behave in restless and impulsive ways, but classify them as showing 'conduct disorder'. By contrast, some US clinicians interpret the definitions of the American Psychiatric Association's Diagnostic and Statistical Manual (1980 & 1987) in such a way that 'conduct disorder' refers primarily to criminal acts and 'hyper-activity' refers to the impulsively antisocial acts that make up most of the disruptive behaviour of children: the result is that hyperactivity is the commonest psychiatric diagnosis of childhood, and stimulants are the commonest treatment. Some European clinicians reject hyperactivity but prefer to diagnose 'minimal brain dysfunction'—for which hyperactive behaviour is one piece of evidence, among others.

All these practices in different countries lead to a common and heterogeneous group of children, differently named but overlapping greatly. Nevertheless, the boundaries are differently defined. Add this to likely differences between centres in referrals and to possible differences between cultures in what adults see as problems in children; and it is small wonder that each psychiatric tradition tends to go its own way. Small wonder, too, if scientific observers see a welter of over-inclusive and incommensurable definitions and conclude that the whole area is best ignored. This cynicism is understandable, but just as wrong as the approaches based on untested theories. They are all too undiscriminating.

Our own investigations (and those of colleagues) were reported in a recent book (Taylor, 1986). We tried to re-examine the status of the condition from a standpoint of scepticism about the neurological theories that gave it birth. We concluded that much of hyperactive behaviour—at least as rated by adults—is situationally specific but aetiologically non-specific. Accordingly, we rejected a narrow medical model of common behaviour problems. We also concluded that a restricted and limited idea of hyperkinetic disorder is a valid concept for studying behaviour.

This concept of hyperkinetic disorder was based primarily upon clinical research. Its characteristics and associations in several studies included: pervasiveness and severity of hyperactivity; a lowered IQ, and, after allowing for that, a poor performance on psychological tests of attention; motor clumsiness; a high frequency of language impairment and other forms of developmental delay; an early onset, nearly always in the years before entry to school; accident proneness and poor relationships with peers; a large response to stimulant medication; and an adverse outcome in terms of psychosocial adjustment during adolescence. Associations with social class and family life were less clear in the clinical studies; but re-analysis of the Isle of Wight population studies suggested that pervasive hyperactivity was linked to lower social class.

Epidemiological investigation is a necessary step in this line of research. Most of the associations have now been replicated in repeated clinical investigations, and they suggest theories of how hyperactivity develops. But, it is essential to know whether the concept of hyperkinesis is still valid and its associations still hold in a population-based series. Referral bias can be a powerful generator of apparent findings in clinic populations. Indeed, differential referral rates could account for the differential associations of hyperactivity and conduct disorder with other disabilities (such as language delay and failures of relationships) that are themselves grounds for referral.

An epidemiological survey is required for immediately practical purposes as well. The prevalence of disorder and the need for services are not clear, and survey information is required to help in planning. For the future, it will be important to be able to study geographical variations in prevalence and compare rates of disorder in different populations. It is therefore desirable to establish a suitable scheme for case finding. This would make possible the classic study designs of epidemiology, in which differences in rates in different populations are used to infer the causes of disorder, and cross-cultural designs are used to seek factors that modify the course and later development of children with hyperactivity.

Epidemiological studies of hyperactivity

Previous studies on the epidemiology of hyperactivity have encountered substantial difficulties in defining cases (Taylor, 1987). In fact, comparison across studies is virtually impossible. There is no single correct method, and various strategies have been adopted.

One approach is to seek nominations, by parents, teachers, or physicians, of children who have been considered to be hyperactive (Bosco & Robin, 1980, Lambert *et al.*, 1978). This gives an idea of the numbers who have been treated: between 10 and 15 per 1000 children had in practice received a medical diagnosis, over an unspecified period. Obviously, however, this is no guide to how many should be recognized, and will reflect primarily the local availability of services and the views of physicians. The 10-year period prevalence rate from a British case register was much less: about 1 in 2000 of normally intelligent children (Taylor & Sandberg, 1984). Clinical diagnosis using DSM–III and ICD-9 categories is too unreliable for a sound study (Prendergast *et al.*, 1988).

Another approach (the commonest) is to survey with parent and teacher rating-scales (e.g. Sandberg *et al.*, 1980, Schachar *et al.*, 1981, McGee *et al.*, 1985). These have the merit of being explicit and convenient measures. Typically they lead to a dimensional scoring system, and an arbitrary cut-off point is taken to define cases. They have serious weaknesses as the sole measures of disorder, and studies based solely upon them have not generated clear or specific findings (Taylor, 1988). They are not sufficient for the definition of an individual case, nor for comparisons between cultures, but can usefully define groups for comparison within a single study when numbers are reasonably large.

The uncertainties about how to define cases from rating scales have given rise to a few investigations with more detailed methods. Boyle and co-workers (1987) took a subsample of the population that they had screened, and obtained clinical DSM–III diagnoses upon them. The cut-off point on the rating scales was then chosen, to make the false positives approximately equal to the false negatives. This method is useful for obtaining a prevalence rate; but could give rise to a high misclassification rate and therefore to uncertainty about the associations of disorders defined by rating scales. (There are also, of course, the difficulties arising from the low reliability of clinical diagnosis and the over-inclusive nature of the DSM–III diagnosis of ADDH.)

Two-stage designs were adopted by Rutter, Tizard, & Whitmore (1970) and Vikan (1985); both of whom screened with teacher and parent rating-scales and examined selected groups with detailed interviews with parents and children. Using the ICD-9 concept of hyperkinetic syndrome, they found very low numbers of cases (respectively, two and three!), so no stable prevalence rate could be given. Shen and co-workers (1985) had a superficially similar design in a study in Beijing: teacher questionnaires identified 161 out of 2770 schoolchildren as hyperactive; the 161 underwent a child psychiatrist's diagnostic evaluation which diagnosed all but one of them as showing ADDH!

With these, more than fiftyfold, disparities in estimating rate it is obvious that some advance in case identification is required. A Swedish study by Gillberg and co-workers (1983) is noteworthy, even though it was directed at minimal brain dysfunction (MBD) rather than hyperactivity. A questionnaire screen for motor and perceptual impairments was followed by a detailed assessment of selected groups using multiple measures and explicit cut-offs. Their tentative point-prevalence rate for hyperkinetic syndrome (30 per 1000) was the same as that suggested for ADDH in the DSM–III diagnostic manual (American Psychiatric Association, 1980).

What is a case?

So far no epidemiological study of childhood hyperactivity, or any of the diagnostic categories based upon it, has been able to adopt reliable and validated definitions, to identify sufficient numbers of representative cases to examine associations, and to compare the resulting groups with other types of disorder. It is worth a brief summary of the obstacles in the way of developing a good answer to the question 'What is a case?'

1. The behaviours that are supposed to characterize the disorder—inattention, overactivity, and impulsiveness—have several different meanings that can be confused, so there is ambiguity about what is meant by the ratings. 'Over-activity' as rated may refer to an excess of movements or to a qualitative change in the patterning of activity. We have aimed to measure both to see which is closest to the rated problem.

'Impulsiveness' is a particular problem because it can mean an over-rapid tempo (which is obviously similar to hyperactivity), or an over-dependence on immediate gratification (which is a theory about cause), or an acting-out of antisocial impulses (which is part of conduct disorder). Accordingly, we have based our definitions of hyperactivity solely upon the behavioural styles of inattentiveness and restlessness. Impulsiveness is still a potentially interesting explanation of the disorder, but is not yet part of its definition. Similarly, we have regarded inattentiveness as a lack of task orientation in behaviour. Many psychological processes contribute to 'attention' (Taylor, 1980) and their relevance to childhood hyperactivity is still unclear. We have based our definitions upon the behaviour, and treated performance upon tests as a dependent variable to study.

2. Inattentiveness and over-activity each has many causes. Over-activity is a feature of anxious, fidgety children; of children who are depressed and agitated; of mania; of autism; and is seen in some children whose development is otherwise normal. We have also seen children referred clinically with the complaint of over-activity, for whom the underlying pathology was in fact not a hyperkinetic disorder, however caused, but rather Sydenham's chorea, schizophrenia, obsessional neurosis, Tourette's syndrome, ataxic cerebral palsy, or disintegrative disorder. 'Inattentive behaviour' is a non-specific complaint about most children with a psychiatric disorder of any type.

We have started from the position of defining the combination of inattentiveness with over-activity as hyperactivity, regardless of underlying cause. We have therefore gone on to study its overlap with other types of disordered behaviour (Chapter 2); and to define hyperkinetic disorder in the light of those findings (Chapters 3 and 7). The likely causes of the disorder then become dependent variables to be studied, not part of the initial definition.

We have also identified children with inattentiveness, but who are not overactive, to determine whether they do indeed have the same kind of problem as the hyperactive—as is implied by those definitions in which attention deficit is the primary pathology. Further, we have used experimental tests of processes related to attention in order to determine whether 'attention deficit' is a misnomer or whether children with inattentive behaviour also have an impairment of the central processes recognized as attention by psychologists.

3. Another difficulty for case definition is that there is no clear dividing line between normal and abnormal levels of activity and attention. The range of 'normal variation' is very wide; though even this remark begs the question of the sources of that variation and its acceptable limits.

The problem is substantial, though of course it is not confined to hyperactivity. Many important biological variables, such as blood pressure, are distributed continuously in the population and arbitrary levels of case definition have been required in the early stages of their study. Ultimately, the answer will depend upon studying different levels of severity to determine the level at which there is significant handicap. We have made a start on this by comparing different levels of case definition (in Chapter 7). However, the full answer will require longer-term follow-up studies and cannot be given yet.

In the meantime, we have tried to avoid arbitrary choices of cut-off points. It is quite possible, for example, that different degrees of over-activity and inattention have different implications. It has been suggested for more than fifty years that social and familial factors dominate the aetiology at less severe levels of hyperactivity, while biological factors are significant at extreme levels (Levin, 1938). If so, the level taken will determine findings.

We have also tried to avoid total reliance upon the diagnostic judgement of a trained psychiatrist, even though this has been useful in investigations of conduct and emotional disorder (Rutter, Tizard, & Whitmore, 1970). The unreliability of the clinical diagnosis of hyperactivity, and the very different numbers identified by different experts have already been stressed. Indeed, the evolution of psychiatric diagnostic schemes has been towards greater explicitness of criteria, so that the DSM–III and DSM–III-R criteria now amount to arbitrary cut-off points on an agreed scale of unknown reliability and validity. There is little to be gained from calibrating one definition of unknown value against another.

Our main approach has been to take provisional cut-offs based upon previous research. The route to this is sketched below, and the exact criteria taken, and their justification, are given in the descriptions of method provided in the main text (Chapters 2, 4, and 7).

4. A linked difficulty in case definition is that of determining the number of dimensions to be used. We have spoken above as though there were a single dimension of hyperactivity, but the proposition needs to be examined (see Chapter 2). Quay (1979) has taken a leading part in arguing that a dimension of 'inattentiveness-immaturity' should be recognized rather than one of hyperactivity. Furthermore, hyperactive behaviour is very often specific to a single situation, such as the school (Sandberg, 1986). Research has not yet shown whether this is a function of error in ratings or of children behaving differently, but we think that both are important. Hyperactivity at home and at school may in effect be separate dimensions, not one. We have found it useful to consider pervasiveness of hyperactivity across situations and persistence over time in setting up groups for study. This may be because pervasiveness and persistence are both indicators that a significant quality of the child is being measured. The implications for this study will be considered in the light of empirical findings about the cross-situationality of behaviour (Chapter 2).

5. Age and gender effects upon activity and attention need to be taken into account in defining abnormality. Unfortunately, there is

too little evidence about the reasons for gender differences or the normal developmental course to be able to do so in a rational fashion. For the present study we have sidestepped the difficulty by focusing investigation upon a narrow age range and confining the study to boys. But the issue will return.

6. Some difficulty arises from uncertainty about the nature of the condition of hyperactivity to be defined. It is possible to consider it as the extreme of a temperamental dimension, as a symptom of many disorders, or as a separate disorder. We shall return to this issue in discussion. However, it need not be seen as a fundamental problem for the design of our study. A dimension can readily be translated into a set of categories. A category will usually be defined on the basis of scores on dimensions—how many, is a matter for judgement. Furthermore, it is not at all essential for a categorical ordering to involve all-or-none dichotomization into the 'presence' or 'absence' of a disorder. It is convenient for analysis to proceed as though there were clear lines of demarcation around a syndrome. It may, however, well be more realistic to conceive of the degree to which a case fits into a category. Cluster analysis, for example, can generate a measure of the similarity of the profile of a single case on various measures to that of a cluster. A diagnosis can also be conceived of as a prototype, so that it is possible to quantify the number of features in common between a case and the prototype of the condition (e.g. Horowitz *et al.*, 1981).

Before such analyses, however, it is necessary to have a clear idea of the nature of the condition. The present research seeks to contribute to that understanding. We have conceived of hyperactivity and defiance as separate dimensions, and chosen categorical groups in order to study the extremes of the range (see Chapter 4). We have thought of hyperkinetic disorder as a potential category of psychiatric disturbance and used several types of information (such as duration of disorder and absence of other disturbances), in addition to dimensional scores, to define it (Chapter 7). It remains to be seen whether this complex idea of a hyperkinetic disorder will in fact add any predictive power to the score on a dimension of hyperactivity.

Design of an epidemiological study

The design of the study followed from its purposes, mentioned above; from the methodological difficulties arising from the uncertainties about defining cases; and from practical constraints. The study had, of course, to be based upon a general population-sampling framework large enough to yield reasonable numbers of the condition even if it turned out to be uncommon, and not skewed by selecting a few schools on the basis of their co-operativeness. Explicit, reliable, and valid criteria of case identification had to be employed. Previous studies had suggested that we would find considerable situation specificity for hyperactive behaviour (Sandberg *et al.*, 1978, Sandberg, 1981), so multiple measures had to be planned, even for an initial screen. Detailed and time-consuming measures were needed for reliable measurement of the factors likely to be important in aetiology: family life, social circumstances and neurodevelopmental status.

These contradictory requirements led to a two-stage design, in which screening measures were applied to a total population sample in a geographically defined area; followed by the application, to selected groups, of detailed measures that had been shown in clinic studies to be reliable ways of identifying a valid group of cases.

Our previous studies had first developed measures that could be applied reliably. Rating scales by teachers were examined in clinic and population samples for their inter-rater agreement, test–retest stability, and factor structure (Taylor & Sandberg, 1984); and for their ability to predict the results of direct observation of children's behaviour and interactions in the classroom (Sandberg, 1986). The Rutter and modified Conners scales were taken as useful. Parents' ratings scales were particularly likely to reflect attitudes and projections by the raters as well as observations and recall; and so they had to be interpreted with appropriate caution.

Direct observation of off-task behaviour relied upon a trained observer watching a child engaged in a task requiring persistence and concentration. The observer used an interval-sampling scheme to code whether specified off-task behaviours had occurred during 10-second periods, and different observers showed good agreement (Sandberg *et al.*, 1978). This observational measure agrees with a psychiatrist's ratings of hyperactive behaviour made during an independent, standardized psychiatric interview (Taylor, 1986, p. 79); ratings which were themselves validated against direct observation of the videotaped interviews (Luk *et al.*, 1987).

The Parental Account of Children's Symptoms (PACS) is a standardized interviewing method intended to provide more detailed and reliable information about home behaviour than can be obtained from parental questionnaires (Taylor *et al.*, 1986*a*). Parental ratings are limited by varying standards for behaviour and varying interpretations of written questions. Nevertheless, parents do of course know far more about their children than anybody else, and their view of children's behaviour is not distorted by the artificial context that is generated when a professional observer enters the home. Accordingly, the PACS is based upon a trained interviewer enquiring into details of behaviour in a whole range of situations. The parent provides the recall of what actually happened; the interviewer provides the judgements about frequency and severity of the behaviour problems described. These judgements are reliable in the sense that there is good inter-rater agreement on the scoring of an interview and adequate test–retest agreement (Taylor *et al.*, 1986*a*). The scales (of hyperactivity, defiance, and emotional disorder) are factorially distinct.

Attention, in the sense of particular kinds of test performance, also needed to be measured in several ways (Taylor, 1980, 1986). Sustained attention, selective attention, efficiency of new learning, and impulsiveness of style all came into the notion. They required a combination of tests, including a Continuous Performance Test of vigilance; a Paired Associate Learning Test requiring persistence and use of strategies for memorizing; the Digit Span; and an improved version of the well-known Matching Familiar Figures Test, of impulsiveness versus reflectiveness. Weighted scores from all the attentional tests were combined to yield a summary scale of 'attentional performance' which had adequate test–retest reliability.

After the measures had been developed and their psychometric properties studied, their predictive validity was examined. Conners' (1969) Classroom Rating Scale was of limited value from this point of view: hyperactivity and defiance could be separated as distinct components but the two were quite highly intercorrelated. The separation could be improved by using true factor scores rather than by summing raw scores on items with unit weight. When this was done, then hyperactivity was more closely associated with developmental delays, and defiance with adverse features of family life. A similar pattern of associations was found for the scales of the PACS interview. The scale of performance on tests of attention was associated with IQ, but even after allowing for this it was significantly associated with measures of hyperactive behaviour—especially with the inattentive component; and both the attention performance scale and the behavioural scales of hyperactivity were predictive of responsiveness to methylphenidate medication. The neurodevelopmental examination gave an overall score of clumsiness, which was higher in children with pervasive hyperactivity. Rutter's teacher and parent rating-scales proved to yield a factor of hyperactive behaviour; and when the score was elevated on both questionnaires it predicted the persistence of disorder from age 9 to 14, as well as the test performance of children.

The exact cut-offs taken on the basis of these studies are given in the main text (Chapters 2, 4, and 7). They still left some uncertainty about the way in which hyperkinetic disorder should be defined.

However, a cluster analytic study had given a cluster of cases, characterized by pervasive hyperactivity and reduced attention, that was both predictive of other clinical features and in keeping with the qualitative findings of our other researches (Taylor *et al.*, 1986*b*). We therefore took the scores on the hyperactivity-inattentiveness factor of the Classroom Rating Scale, the hyperactivity factor of the PACS interview, and the observational measure of off-task behaviour during testing. These measures gave a perfect identification of the cases in the hyperkinetic cluster (almost by definition) and excluded the other clusters of disruptive behaviour. These cut-off scores on these measures were taken as the operational definition of 'severe and pervasive' hyperactivity.

The diagnostic criteria suggested by our research include aspects other than the severity and pervasiveness of hyperactivity and off-task behaviour. The full list requires the presence of all of the following:
1. A pattern of markedly inattentive, restless behaviour—not just antisocial or defiant symptoms.
2. The presence of this pattern to a degree that is unusual for the child's age and level of development.
3. The presence of this pattern in two or more situations (such as home, school, and clinic).
4. Directly observed evidence of inattention, restlessness, or disinhibition.
5. Absence of childhood autism, other childhood psychoses, and affective disorder (such as depression, mania, or anxiety state).
6. Onset before the age of six years, and duration of at least six months.

These criteria are similar to those of ICD-10 (see Chapter 8). Although ICD-10 had not been drafted when this study was started, its definition of hyperkinetic disorder is sufficiently close to the one used here that the final stage of our study can reasonably be seen as an examination of the validity and prevalence of the ICD-10 category of hyperkinetic disorder.

This chapter has indicated the reasons for the study, some problems for research which will be taken up again in the discussion, and the general rationale for the methods used. The following chapters will describe the methods and findings of the different stages of the study; and a final discussion chapter will draw out the general conclusions. The Outline at the beginning of this volume provides a summary overview of the results.

Screening of the population with questionnaires

2 Classification of disruptive behaviours

INTRODUCTION

Chapter 1 raised the issue of whether disruptive behaviour can be usefully classified, and in particular whether hyperactivity can truly be distinguished from the behavioural style of defiance and aggression. The question is not finally settled.

We have therefore tried to confirm or deny the separability of conditions of hyperactivity and conduct disorder, and to clarify the associations of each. This chapter reports the first stage of the inquiry, the questionnaire data on the entire sample: and with these data we address the issue of how far descriptive ratings support the separation of distinct subtypes of disruptive behaviour.

PREVIOUS TAXONOMIC STUDIES OF RATING SCALES

Questionnaire rating scales completed by teachers or parents have often been used before as the raw material for studies of classification of child psychopathology. Indeed, they have been used almost to the exclusion of other measures. Studies employing them have generally taken, in the attempt to determine whether they belong with either of the pure subgroups, one or more of three approaches: exploratory factor analyses to extract dimensions from correlations between rated items of behaviour; comparisons of groups defined by the presence or absence of hyperactivity and aggression; and the description of mixed groups showing both problems.

Factor analytic studies of rating scales

Hinshaw (1987) has made a full review of 60 published studies (1970 to 1986) that have applied factor analysis to rating scales for samples of at least 100 children with average ages between 4.5 and 14.5 years. (The earlier literature is covered in reviews by Achenbach and Edelbrock, 1978, and Quay, 1979.)

In 41 of the 60 studies, there was a factorial distinction between conduct problem/aggression on the one hand and attention deficit/hyperactivity on the other. Of those that did not find such a distinction, several would have been bound to fail since the scales did not include sufficient items relating to motor activity or inattention. Indeed, the preponderance of studies finding the factors to be separate is even more marked in the studies published in the last decade, which have in general been based upon wide and systematic collections of items.

This consensus between studies makes it improbable that the findings are an artefact. Although many individual studies have too small samples for factor solutions to be robust, and although there is a wide range of statistical and sampling techniques, they amount in sum to a persuasive argument. It is possible that they could all share a systematic bias of some kind; for example, the apparent independence of factors could result from a negative halo effect, reflecting the attitudes of raters rather than the behaviour of children; and attempts at replication across cultures should help to clarify this (Luk & Leung, 1989).

A more serious objection comes from the lack of agreement between sources. The correlation between teachers' and parents' ratings of hyperactivity is often smaller than that between two, supposedly orthogonal, subscales from teachers (Sandberg, 1981).

More serious still, there is a very large overlap between factors. Many of the items that define hyperactivity subscales also load highly on factors of conduct disorder—sometimes the same items are included in more than one subscale. The subscales of hyperactivity and conduct disorder are significantly intercorrelated in nearly all the studies that report the degree of association. The correlations are so high (median = 0.56 in Hinshaw's cited review) that the independence of the subscales is in doubt.

The interpretation of factor structure is also far from straightforward. One must take care lest the factors be reified into traits; they are only a summary description of intercorrelations. A pattern of intercorrelations could be due, for instance, to an aspect of severity of disturbance or to distributional properties of items. Before dimensions of disorder can be accepted as valid, they must show predictive and discriminative associations with aetiology or course. This has been much less well examined. Indeed, the studies showing the clearest differential associations—of hyperactivity with developmental delay, and of conduct disorder with psychosocial adversity—are based upon richer and more detailed measures of behaviour than simple rating scales (Loney *et al.*, 1978, Taylor *et al.*, 1986a).

In summary, the factoring studies have given modest support to a partial independence of hyperactivity and conduct disorder; no more. Research therefore needs to develop towards using large, representative samples, taking account of parent and teacher ratings together, examining the extent of the overlap between different types of deviance, and studying the discriminative validity of the types.

Comparisons of distinct groups

Taylor (1988) has reviewed the empirical studies that have attempted to contrast cases assigned by standardized assessment measures to categories of hyperactivity (HA) or conduct disorder (CD). There have been relatively few, not least because of the practical difficulty of finding 'pure' cases of either condition in the clinic-referred groups of children who are usually investigated. In practice, many of the comparisons have been between hyperactive children with greater or lesser degrees of antisocial conduct (e.g. Werry *et al.*, 1987) or else between conduct-disordered children with greater or lesser degrees of hyperactivity (e.g. Sandberg *et al.*, 1978).

Results of the direct comparisons are mixed. In very brief summary, Stewart and co-workers (1981) found HA to be symptomatically distinct from CD in disturbed children at a tertiary referral centre; this was not highly informative, since both symptom scales and diagnostic classification were taken from parental accounts, but helped to confirm the separability of the behaviour patterns. More usefully, the symptomatic distinction was still present at 4-year follow-up (August *et al.*, 1983). Loney and Milich (1982) found HA children to be more active and more off-task than CD when directly observed. Trites and Laprade (1983), reporting a large population survey, duly found separate factors of hyperactivity and conduct disorder, and then found the HA children to be less conduct-disordered than the CD, and the CD to be less hyperactive than the HA. This analysis, of course, merely repeated that there was less than perfect agreement between the two factors. McGee and co-workers (1984) described a smaller group of children, forming a birth cohort, who were examined at 7 years. The HA group was (of course) more hyperactive, more labile and more impulsive than the CD; performance IQ was a little lower in HA, but most measures of background factors did not distinguish groups. Shapiro and Garfinkel (1986) found HA and CD 'cases', derived from a single elementary school, to be indistinguishable in symptomatology assessed by structured interviews.

Other designs are less than conclusive. Werry and co-workers (1987) have reviewed the few studies comparing different DSM–III diagnoses, including ADDH and CD. The lack of substantial differences between them may reflect chiefly the inadequacies of that diagnostic definition and the consequent mixture of behavioural problems in each diagnostic group.

It will be clear that conclusions are impossible in current knowledge. Some quite basic information is missing. We still need to know whether HA and CD groups differ in symptoms other than those that define them, and in total numbers of symptoms. If, for instance, HA were associated with affective symptomatology or an increased severity of disturbance, then any other differences between groups could be the result of greater severity of disturbance rather than the presence of HA. This issue has not been sufficiently raised in the literature, and will therefore be addressed in this chapter.

Status of mixed cases

Most clinically referred cases are 'mixed'. For example, Stewart and co-workers (1981) found 61 per cent of those with hyperactivity to show definite conduct disorder, and 67 per cent of the conduct-disordered to show definite hyperactivity: many more showed less marked mixtures. Even when using subscales supposed to maximize the separation of HA and CD, Loney and Milich (1982) found 75 per cent of their hyperactive boys to be aggressive too. Even in UK clinics, where any diagnosis related to hyperactivity is rare, most children with conduct disorder meet DSM–III criteria for ADDH (Taylor *et al.*, 1986b).

The effect of referral bias probably increases the degree of overlap in clinic cases. The ratio of mixed cases therefore needs to be assessed in population samples too. Trites and Laprade (1983) found 65 per cent of the conduct-disordered to be hyperactive and 64 per cent of the hyperactive to be conduct-disordered. Although they stressed the incompleteness of the overlap, it is in fact similar to that in the tertiary referral centre of Stewart and co-workers (1981).

It should be very useful to determine whether mixed cases belong with the HA or the CD. In other contexts (such as the overlap of diabetes and heart disease) this approach has been a valuable way of understanding aetiology. The view, expressed by Loney and co-workers (1978), that hyperactivity symptoms are primary and conduct disorder secondary is akin to that of Taylor (1986) that hyperactivity is a risk factor for the development of conduct disorder through the mediation of family relationships. Both these notions predict that the mixed group (HA-CD) should evolve from HA and therefore be classified with it rather than with CD.

In spite of the significance of the question, we know little from published studies about the answer, even in terms of behavioural symptoms. Stewart and co-workers (1981) found HA-CD to resemble CD rather than HA on the basis of their symptomatology. On the other hand, August and Stewart (1982) presented another analysis of the same (or very similar) group of children, in which the requirement of pervasiveness across measures was added to the definition of hyperactivity. In these circumstances, there was little difference between HA and HA-CD. McGee and co-workers (1984) found that HA-CD was the only group with high prevalence of specific reading retardation, and its outcome was particularly unfavourable. Like HA, its mean performance IQ was low; like CD, its rate of single-parent or broken homes was high. These findings would be in keeping with a dimensional classification, in which children with both problems have the associations of both; but could equally well imply that HA is an index of the severity of CD.

The lack of secure information on this question leads us to report the nosological position of the HA-CD group in the present study. There is obviously no point in comparing the HA-CD with the CD on hyperactivity symptoms, for they are bound to be different by definition. However, a dimensional model predicts that the HA-CD will be similar to the HA in hyperactivity symptoms and to the CD in antisocial; while the profiles of symptoms should differ between groups in a categorical model.

This chapter needs to answer several questions: Are hyperactivity and conduct disorder factorially separate in this population, using this scale? What is the relationship between them, and are they dependent on the situation in which they are measured? How are scores on these scales distributed in the population? Do HA and CD groups differ in total numbers of symptoms and in symptoms of emotional disorder? Can the HA-CD group be classified with either the HA or the CD?

These subsidiary questions all need to be asked in order to examine our major issue, of whether hyperactivity and conduct disorder are distinct types of problem.

METHODS

Subjects

The subjects for this study were all the boys, aged between 6 years 0 months and 7 years 11 months on January 1st, who were at schools in the London borough of Newham.

Newham is a local government area in East London, with a mixture of qualities of inner and outer city areas. Education and community health services are organized on the same borough-wide basis. It is a relatively poor borough, with both a traditional local identity and a high concentration of ethnic minority groups (especially immigrants from the Indian subcontinent). Unemployment rates are high, and there has been a collapse of the major industry of the docks followed by some redevelopment of the area. It should be seen as a stressed urban community, in which prevalence rates of childhood disorder are likely to be higher than in rural or more advantaged areas.

Measures

Rutter's B(2) questionnaire was completed by teachers, and his A(2) questionnaire by parents. These are rating scales in which a wide range of items about perceived problems are rated on a 3-point scale (Rutter, Tizard, & Whitmore, 1970). They have been used in many epidemiological surveys, with acceptable reliability. They have the particular merit for our purpose that cut-off scores defining the presence of deviance are not arbitrary but have been validated against standardized interview measures and psychiatric diagnosis (Rutter, Tizard, & Whitmore, 1970) or by the identification of different subgroups with predictive and discriminative validity (Schachar *et al.*, 1981).

On each scale, items are summed with unit weight to yield an overall measure of deviance. Those with a score of 9 or greater on the teacher scale, or 13 or more on the parent scale, are classed as deviant. The number of symptoms on the 'conduct disorder' (aggressive, disobedient, antisocial) and 'neurotic' (anxious, miserable, phobic, somatic) subscales are then calculated, and the case assigned to a category of conduct disorder or emotional disorder, according to which type of symptom predominates. A 'hyperactivity' subscale is also calculated on the basis of 3 items from each scale (restless, fidgety, not settling to activities), which are not part of the conduct disorder or neurotic subscales. Children with a score of 3 or more on both scales are classed as 'pervasively hyperactive' (Schachar *et al.*, 1981).

The three items of the hyperactivity subscale were rather few for a reliable scale, and they stress the motor aspect of the construct rather than the inattentiveness. Accordingly, seven items from the Conners teacher rating-scale (Conners, 1969) were included, being the items that defined the factor of inattentive-restlessness reported in our previous epidemiological work on the scale (Taylor & Sandberg, 1984). The items are shown in Table 1—fidgeting, making odd noises, restlessness, inattention, failure to finish things, disturbing others, poor co-ordination. Their inclusion allowed us to determine whether the subscales from the two questionnaires were comparable.

Procedure

All the schools in the borough were identified from central lists in the Education Department, and all were approached to take part in the survey, except for units dealing only with children who had severe learning disabilities. We were able to obtain names and dates of birth from all the school registers, and so to identify the subjects for the study. The class teachers of the subjects were then asked to complete a rating scale for each of them.

The parents of all the subjects were sent, through the schools, an explanation of the study and a rating scale to complete and return. When scales were not returned at this stage, two more approaches were made by direct mail; and, for some of those who had still not responded, a personal visit was made to explain the study.

Confidentiality was rigorously maintained throughout. Feedback was given to individual schools on their rates of deviance and the rate in the borough as a whole, but no information was released on individual children.

Analysis

Parent and teacher rating-scales were first factor-analysed separately, in order to examine the separability of dimensions of behaviour problems. Principal component analysis was performed, with rotation of factors to varimax criteria. When comparable dimensions were found on both scales, the agreement between teacher and parent ratings was examined to determine how far each dimension could be seen as a reliably elicited trait of the child. The distributions of scores on the scales of hyperactivity and conduct disorder were examined for any evidence of bimodality. Only those children scoring above the cut-off on one or both of these scales were admitted into the next stage of the analysis. Three deviant groups could then be established: conduct disorder without hyperactivity (CD), hyperactivity without conduct disorder (HA), and a mixed group showing both kinds of problem (HA-CD). The 'pure' groups (HA and CD) were compared on the basis of total symptom scores and emotional disorder scores. The relationship of the mixed cases (HA-CD) to the other two groups was examined in two ways. First, those with hyperactivity and conduct disorder (HA-CD) were compared to those with conduct disorder only (CD) according to their rates of different conduct disorder items;

and to the HA for their rates of hyperactivity symptoms. Second, a discriminant function analysis was employed, using all items rated with equal weight in a stepwise analysis. This began with the construction of a single discriminant function providing optimal separation of an HA group (with pervasive hyperactivity, but not meeting criteria for CD) and a CD group (which was classified as conduct disorder but scored 2 or less on both hyperactivity subscales). Next, the same function was used to classify the HA-CD group into one or other of the 'pure' groups. Finally, the distribution of each of the three groups on the discriminant function was plotted.

All statistical procedures were carried out as maintained in the SPSS-X program package (SPSS, 1983).

Results

There were 3215 boys on school rolls whose birthdates were within the range of the study (plus another 8 who were at units for children with severe learning delays and were therefore excluded from this stage of the survey). Teacher rating scales were obtained for 3176 (98.8 per cent); and, of those, parent rating scales were completed by 2462 (78.2 per cent). Another 6.9 per cent of parents returned their forms blank or otherwise indicated that they wished not to participate. Of the non-respondents, 25 were approached by personal contact to ask their reasons for not participating; the answers given were predominantly that forms had been mislaid.

The children of parents who did not return scales were rated by their teachers as slightly more deviant than those of parents who replied. The total score on the Rutter B(2) scale had a mean of 6.9 for the children of non-respondents and 5.8 for respondents ($t = 4.2$, df $= 3174$, $p < 0.01$). The equivalent figures for the hyperactivity subscale were 1.6 and 1.3 ($t = 3.3$, $p < 0.01$); for the conduct disorder subscale 1.6 and 1.3 ($t = 3.1$, $p < 0.01$); and for the neurotic subscale 1.3 and 1.2 ($t = 2.5$, $p < 0.05$).

The components emerging from principal component analysis of the teacher rating scale are set out in Table 2.1. The five factors accounted for 60 per cent of the variance in the unrotated matrix. When Conners and Rutter scales were analysed together, then components emerged in the order shown in the table, with 'hyperactive' accounting for most of the variance and 'antisocial' least. When the Rutter scale was analysed separately, the same components emerged with the same scale items loading on them. However, the 'conduct disorder' component was now the largest, and 'hyperactive' emerged in third place after 'neurotic'. Two items ('lies' and 'fails to finish things')

Table 2.1. *Principal components of the teacher rating scales*

	Hyperactive	Conduct disorder	Neurotic	Withdrawn	Antisocial
Rutter B(2) Scale	Very restless Squirmy, fidgety Cannot settle to anything	Frequently fights Irritable, touchy Disobedient Lies Resentful, aggressive Bullies	Worries Miserable Twitches Fearful Fussy Tearful at school	Solitary Unresponsive	Destructive Lies Steals
Conners Scale	Constantly fidgeting Makes odd noises Restless, over-active Inattentive, easily distracted Fails to finish things Disturbs other children			Co-ordination poor Lacks leadership Fails to finish things	
Per cent of variance explained:	30.6%	10.4%	6.4%	4.9%	3.5%

The table shows all items with loadings of 0.40 or greater on each component; the components shown are those emerging when the Rutter and Conners items are analysed together

Table 2.2. *Principal components of the Rutter A(2) parent rating scale*

	Conduct disorder	Hyperactive	Neurotic	Antisocial	Somatic
Rated item	Temper tantrums Frequently fights Irritable Disobedient Bullies	Very restless Squirmy Cannot settle	Worries Solitary Miserable Fearful	Destructive Lies Steals	Headaches Stomach-aches
Per cent of variance explained:	17.3%	6.0%	4.4%	4.3%	4.0%

The table shows all items with loadings of 0.40 or greater on each component loaded on two components and are shown on both. Different criteria of rotation (oblimin with values of delta varying between $+0.5$ and -0.5) yielded the same components and items loading on them, but necessarily varied in values of individual item loadings.

Table 2.2 shows the equivalent results from principal component analysis of the A(2) scale completed by parents. On this instrument, the five components initially extracted accounted for only 40 per cent of the variance, and there was a large number of components with eigenvalues less than unity, each having significant loading from only one item.

The analysis therefore supported the distinction between behavioural dimensions of hyperactivity and conduct disorder; and indicated similarity of the hyperactivity measures of the Rutter and Conners teacher rating scales.

The subscales of conduct disorder and hyperactivity were then correlated for teacher and parent ratings: results appear in Table 2.3.

All the correlations are significant, but those between hyperactivity at home and school, and between conduct disorder at home and school, are no higher than 0.3. Figures 2.1 and 2.2 show the distribution of scores on each subscale: none of them shows any obvious sign of more than one population of cases. However, though a dimensional ordering seems appropriate, a small but distinct category at the extreme of the scales would not necessarily have been evident.

Accordingly cut-offs for the definition of the groups to be studied were taken on the basis of levels validated in previous surveys (see 'Measures').

In Table 2.4, the resulting deviant groups are compared with one another and with the non-deviant children—i.e. all those for whom ratings were available, who did not meet criteria for hyperactivity,

Table 2.3. *Behaviour according to different raters*

		Teacher Conduct Disorder	Parent Hyperactivity	Conduct Disorder
Teacher				
	Hyperactivity	0.61	0.24	0.21
	Conduct disorder		0.19	0.30
Parent	Hyperactivity			0.43

The table shows product-moment correlations between subscales of Rutter's A(2) and B(2) rating instruments, carried out on 2462 6- and 7-year-old boys. All correlations are statistically significant ($p < 0.01$)

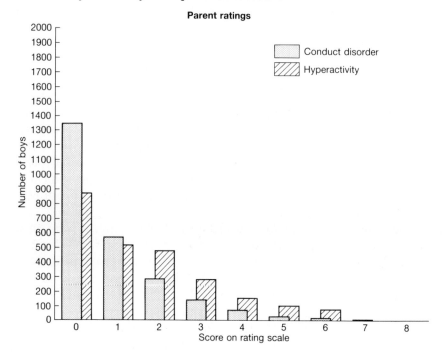

Fig. 2.1. Distribution of scores on conduct disorder and hyperactivity scales of the Rutter parents' questionnaire.

and whose total scores were below the cut-offs for deviance. Despite the intercorrelation of subscales it was possible to separate pure and mixed subtypes of disorder, as shown. The exclusive groups of pure hyperactivity (HA) and pure conduct disorder (CD) are similar to one another in severity of disorder (indexed by total number of symptoms) and have fewer symptoms than the mixed group.

The neurotic subscale of the questionnaires was significantly elevated above non-deviant controls only in the HA group. For the teachers' scale, all the items comprising the neurotic subscale were more commonly present in the HA. For the parents' neurotic subscale, only the item 'is there any sleeping difficulty?' was more often answered affirmatively. The hyperactive children who scored 4 or more on the teachers' neurotic subscale (N = 31) were similar to the other hyperactive children (N = 190) on all the items of hyperactive and antisocial behaviour; they were more likely to be unpopular or isolated with peers; by comparison with the non-hyperactive children scoring high on the neurotic scale (N = 285) they were similar in most of the items of emotional disorder but more likely to show the definite presence of 'misery' (48 per cent v. 23 per cent, $\chi^2 = 8.1$, df = 1, p < 0.01).

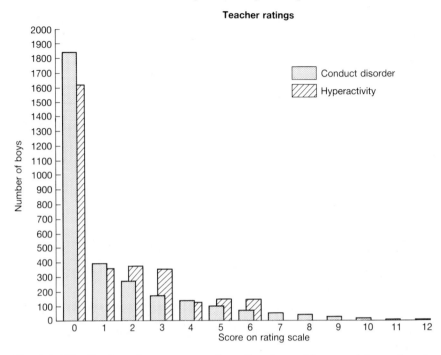

Fig. 2.2. Distribution of scores on conduct disorder and hyperactivity scales of the Rutter teachers' questionnaire.

The mixed group (HA-CD) was compared with the CD group for symptoms of conduct disorder (Table 2.5) and with the HA group for symptoms of hyperactivity (Table 2.6).

All the symptoms of conduct disorder (except truancy, a rare problem in this age range), were more common in HA-CD than in CD; all the classroom symptoms of hyperactivity were more common in HA-CD than in HA, and parents' rating of poor attention was also commoner in HA-CD. It seemed that the increased number of symptoms in the HA-CD group was not only a matter of the definition requiring more symptoms, but also reflected the fact that HA-CD were more severely affected.

The position of the mixed group was examined with discriminant function analysis (see 'Analysis'). The analysis was carried out only on those cases who showed hyperactivity and/or conduct disorder by the definitions already set out, and children scoring 4 or more on either of the 'neurotic' subscales were excluded. Figure 2.3 shows the distribution of cases on the discriminant function score. It naturally gives a complete separation of the HA and CD groups, for it contains the symptoms that defined the groups. Most of the HA-CD cases are

Table 2.4. *Ratings of other symptoms in questionnaire—defined groups*

	Group			
	Pure conduct disorder	Pure hyperactivity	Mixed HA/CD	Control
	N = 349	N = 91	N = 128	N = 1865
	Mean (S.D.)	Mean (S.D.)	Mean (S.D.)	Mean (S.D.)
Teacher rating				
Emotional disorder*	1.1 (1.4)	2.0 (2.1)**	1.2 (1.5)	1.1 (1.6)
Total score*	11.6 (6.6)	10.4 (6.0)	16.2 (7.2)**	3.9 (4.1)**
Parent rating				
Emotional disorder*	1.5 (1.3)	2.5 (1.9)**	2.0 (1.5)	1.7 (1.5)
Total score*	11.8 (6.9)	14.0 (6.9)	17.2 (6.6)	7.8 (5.3)**

*Analysis of variance shows differences between groups, $p < 0.01$.
**Pairwise testing by the method of least-significant differences indicates a significant difference from all the other 3 groups, $p < 0.05$.
Note 29 cases were excluded because teacher or parent had not rated some of the emotional disorder items

intermediate between the two pure groups, as would be expected from a dimensional arrangement. The HA-CD group is similar to the HA: 78 per cent of the mixed group were classified as pure HA by the discriminant function; only 22 per cent were placed with conduct disorder.

DISCUSSION

The return rate was good for a survey of this type, but leaves some scope for sampling errors. The prevalence of hyperactive behaviour is likely to be slightly underestimated, given the slightly higher teacher rating scores of those whose parents did not respond. Further, it was based upon those at school in the area rather than those resident in it. Though the two are very similar they are not identical. In particular, we were not able to identify children who lived in the borough but went to fee-paying independent schools outside it. They are unlikely to be numerous: the boys were too young for entry to most private schools. No boys in the age range had been placed outside the borough by the education authority. We consider that the sample of boys studied was adequately representative of boys in the area; and in particular

Table 2.5. *Percentages of children in diagnostic groups showing definite presence of conduct disorder and antisocial symptoms*

	Pure conduct disorder	Mixed hyperactivity and conduct disorder	
	N = 349	N = 128	
	%	%	
Teacher ratings			
Truanting	0.3	4.8	*
Destructive	6.6	18.1	*
Quarrelsome, fighting	25.9	33.9	*
Disobedient	17.8	40.6	*
Lying	10.3	25.2	*
Stealing	6.3	10.2	NS
Resentful	14.4	24.2	*
Bullying	15.5	18.1	NS
Sullen	5.2	12.5	*
Parent ratings			
Temper tantrums	21.4	31.5	*
Truanting	0.6	0	NS
Destructive	11.1	18.3	*
Quarrelsome, fighting	12.9	22.7	*
Disobedient	18.3	41.7	*
Lying	11.0	22.0	*
Bullying	6.9	7.1	NS
Stealing	1.2	4.7	*

*Differences between groups statistically significant by chi-square, $p < 0.05$

we do not think that children with hyperactive behaviour in the community would have gone preferentially to school elsewhere.

The overlap between ratings within each source, and the disagreement between sources emphasized the unsuitability of rating scales for the purpose of diagnosing individual children. The lack of agreement across differing sources of ratings is in accord with other studies based upon questionnaires. Some of it probably stems from error in each measure, for many other factors beside the behaviour of the child enter into a rating. Some of the disagreement may also come from situation specificity in children's behaviour. All children are, of course, sensitive to the contingencies and expectations of their behaviour in different contexts. Some of the disagreement will come because the two measures

Table 2.6. *Percentages of children in diagnostic groups showing definite presence of symptoms of hyperactivity*

	Pure hyperactivity	Mixed hyperactivity and conduct disorder	
	N = 91	N = 128	
	%	%	
Rutter classroom scale ('definite' rating)			
Restless	44.0	65.6	*
Fidgety	36.3	57.8	*
Cannot settle	19.8	46.9	*
Conners classroom scale ('very much' rating)			
Fidgety	20.9	42.2	*
Restless	15.4	35.2	*
Inattentive	20.9	51.6	*
Short attention span	16.5	40.6	*
Disturbs others	9.9	30.5	*
Makes odd noises	5.5	14.2	*
Clumsy	16.5	13.3	NS
Rutter parental scale ('definite' rating)			
Restless	63.7	64.6	NS
Fidgety	37.5	35.8	NS
Cannot settle	29.7	43.3	*

*Differences between groups statistically significant by chi-square, $p < 0.05$

are made at slightly different times; and children's behaviour varies from time to time.

There are several possible ways of proceeding with classification in the face of this difficulty. The simplest is to avoid assumptions that parent and teacher ratings are measuring the same thing, and to confine analyses and discourse to single sources of data. This is the wisest course when research begins, as it allows for examining the question of whether behaviour in different settings follows the same rules. It is the course that we have followed in some previous investigations, in which we have pointed out the lack of discriminative validity of ratings of hyperactivity taken from any one source. However, even if we conclude that each situation must be considered separately, we still aim to develop a typology of problems and therefore a scheme that synthesises different types of information (Taylor *et al.*, 1986*b*).

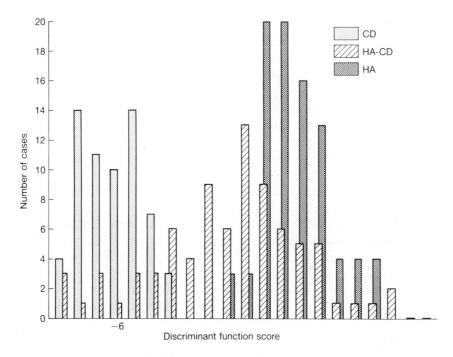

Fig. 2.3. Distribution of scores of hyperactive and conduct-disordered boys on a discriminant function separating the hyperactive from conduct-disordered.

One possible way of integrating competing sources of information is to give priority to one source because it is considered the best guide. This general approach is advocated by DSM–III, which specifies that priority should be given to teachers' accounts over parental accounts and clinical observation (APA, 1980). This is a clear rule, but entails all the possible pitfalls of an arbitrary classification procedure. It ignores the possibility that the degree of situation specificity is a key diagnostic consideration.

Another approach to case recognition is the blending together of different source ratings into a single composite scale. At its simplest this involves treating a number of source ratings as items for an overall scale (e.g. Stewart *et al.*, 1981). At a more complex level, Fergusson and co-workers (1987) have shown how repeated measures by different raters can allow for structural modelling equations yielding a latent dimension of conduct disorder. This latent dimension then gives a different pattern of intercorrelations with external validating variables (higher for some, lower for others) from that of the parent and teacher rating-scale scores. It is not yet clear how far the assumptions involved in such a procedure are justified, but the approach is likely to be fruitful.

We commented in Chapter 1 on the distinction between pervasive and situational hyperactivity, and accordingly have selected subjects for whom hyperactivity is manifest in more than one source. This is similar in effect to a latent class approach, but comes from a somewhat different theoretical basis. We have argued that persistence and pervasiveness of hyperactivity are important because they point to a quality of the child rather than the current environment (Sandberg, 1986, Taylor, 1986).

The high intercorrelations between subscales emphasize the confounding of the measures, for these as for other questionnaires. The subscales are made by summing scores on only those items considered to load on the scales and giving each item unit weight. It is of course possible to achieve better separation by estimating the factor scores and we have used this technique previously in clinic surveys (Taylor *et al.*, 1986*a*). However, every study of a questionnaire will of course yield somewhat different sets of item loadings, so there would be a danger of non-comparability between investigations. Furthermore, even factor scores are significantly intercorrelated (due to correlated error variance). The partial confounding of the measures makes it necessary for research to select relatively pure groups of HA and CD, and to add other assessment instruments.

The continuous distribution of cases along the subscales of hyperactivity and conduct disorder argue for a dimensional rather than a categorical typology. Cases at the extreme of the dimension may still reflect the operation of different factors from those in an intermediate range.

Comparisons of groups

The pure groups of HA and CD are not different in total symptom scores so are not interpretable as different levels of severity. However, the HA group's high scores on symptoms of emotional disorder point to a possible confounding factor. The HA group cannot be simply interpreted as a form of emotional disorder, for the factorial separation of hyperactivity and emotional disorder was even clearer than that from conduct disorder. In the case of the parents' ratings, a ready explanation is at hand. The high scores of the HA group are due to a single item of 'sleeping problem'. There is no reason to assume a general over-emotionality in this context, and it seems plausible that restlessness at night should be part of the complex of behaviours of hyperactivity.

Classroom ratings, however, tell a different story. Nearly all emotional problems are commoner in the HA. The finding raises the possibility that some hyperactive behaviour in boys at school results from a more widespread emotional disturbance. It emphasizes that studies of hyperactivity, that define their cases by a score above cut-off on a measure of hyperactive behaviour, run the risk of spurious findings attributable to other clinical features. It will be desirable for future studies of hyperactivity to exclude those with overtly emotional symptoms, and we have followed this course in subsequent stages of the present study.

The mixed group of cases (HA-CD) is more severely disturbed than the other two groups. Their pattern of symptoms places them, as a group, intermediate between HA and CD. Rather more are classified with the HA (as hypothesized in the Introduction), but their distribution is compatible with a dimensional ordering. However, the mixed group is not sufficiently described simply by saying they show two problems. The mixed HA-CD cases are unequivocally more conduct-disordered than the CD and more hyperactive than the HA. This calls into question the frequent research design that compares hyperactive children (many of whom have conduct problems) with a non-hyperactive control group (whether or not the controls have conduct problems). Any positive findings may not be specific to hyperactivity but could also result from hyperactivity being a marker for increased severity of conduct disorder.

The findings are in keeping with a dimensional classification of disorders, and raise several problems for any categorical scheme. They do not exclude the possibilities that there is a single large category of disorder, altering in presentation but not in fundamental nature; or that there is a number of categorically separate but uncommon disorders. They indicate the possibility and the importance of selecting pure groups for detailed study; and this process will be described in Chapters 4 to 6.

3 Mixed disorders of conduct and emotion

INTRODUCTION

The common psychiatric disorders of childhood are usually classified into two main types: conduct and emotional disorder. The distinction between them is supported by clinical consensus, factor analyses of rating scales, and distinct patterns of association with other variables (Rutter and Garmezy, 1983). The previous chapter has suggested the addition of another type, hyperactivity, not the replacement of either.

Nevertheless, many children show the problems of both emotional and conduct disorders. These 'mixed cases' are common in clinical practice, and their position in classification is not agreed. Some authorities recommend that mixed cases be included with conduct disorder; and the 9th revision of the International Classification of Diseases contains 'mixed disorders of conduct and emotions' as a subgroup of 'disorders of conduct specific to childhood' (WHO, 1978). Others prefer that both aspects of mixed cases should be separately recognized; both DSM–III and DSM–III-R encourage the making of multiple diagnoses when several problems are present (APA, 1980, 1987). Others again advocate that mixed cases should be allocated either to a conduct or an emotional disorder group on the basis of which symptoms predominate: and this determines the scoring of Rutter's A and B scales described below.

The uncertainty creates difficulties for both the definition of cases for research study and the clinical understanding of common problems. This chapter therefore considers whether the disorders are naturally grouped in a way that illuminates the nosological status of children with mixed disorders of conduct and emotions.

METHODS

Subjects

As in the previous chapter, the findings are based upon the community population of 6- and 7-year-old boys, using parent and teacher rating scales. The 2462 boys for whom teacher and parent ratings were available are the subjects of this report.

Measures

Rutter's B(2) questionnaire was completed by teachers, and his A(2) questionnaire by parents. In these questionnaire scales, a broad range of items about perceived problems are each rated on a 3-point scale ('does not apply', 'applies somewhat' and 'certainly applies'). A subscale of 'conduct disorder' is then constructed by summing the scores for those items that deal with antisocial or aggressive symptomatology; and one of 'emotional disorder' from the items of overt affective disturbance.

   The usual scoring scheme involves first the calculation of whether a child's behaviour is deviant (i.e. has a total score greater than a cut-off) and then the subclassification into a group of conduct or emotional disorders on the basis of which subscale score is greater. We did not adopt this system for the purposes of this chapter, as it prejudged the issues of classification that we wished to address. The analysis is based throughout on the two subscale scores of 'emotional' and 'conduct' disorder, considered as separate dimensions.

Analysis

The distribution of scores on each subscale was plotted. Cases were then selected for further analysis if they had a score of 5 or greater on at least one of these scales. Those with scores of 5 or more on either of the conduct disorder scales, but not meeting criteria for emotional disorder, were classified as 'pure conduct disorder' (CD). Those with 5 or more on either emotional disorder but not meeting criteria for conduct disorder were classed as 'pure emotional disorder' (ED). Those with 5 or more on one (or both) of the conduct disorder *and* one (or both) of the emotional disorder scales were classed as 'mixed disorder'.

   The frequency of each rated item was then examined. The items of the emotional disorder scale were compared between the ED and the mixed groups; those of the conduct disorder scale between the CD and the mixed groups. The purpose was to determine whether the profile of symptoms of emotional or conduct disorder remained the same when combined with the other condition.

   A discriminant function analysis was then carried out, and a single function constructed, using all items in a stepwise procedure, to give maximum separation of CD and ED groups. For this purpose, the CD group had the extra requirement of scoring zero on both subscales of emotional disorder, and the ED group had to score zero on both subscales of conduct disorder. The same function was then used to classify the mixed cases.

RESULTS

Factor analysis of the questionnaires in the total sample had previously indicated separate dimensions of conduct disorder and emotional disorder in this large sample (see Chapter 2). The distribution of scores on the parent and teacher subscales of emotional disorder was continuous, with progressively fewer children at successive higher scores.

Table 3.1. *Symptoms of emotional disturbance in contrasting groups of emotional disorder*

	Emotional disorder		Mixed emotional and conduct disorder		
	N = 205		N = 32		
	No. with symptom	%	No. with symptom	%	
Teacher rating					
Worried	98	(48)	16	(50)	
Miserable	37	(18)	16	(50)	*
Fearful	81	(40)	9	(28)	
Fussy, over-particular	14	(7)	7	(22)	*
Apathetic	18	(9)	5	(16)	
Tears on arrival at school	26	(13)	6	(19)	
Parent rating					
Worried	76	(37)	13	(41)	
Miserable	12	(6)	7	(22)	*
Fearful	54	(27)	7	(22)	
Fussy, over-particular	17	(8)	8	(25)	*
Tears on arrival at school	15	(7)	4	(13)	

The table shows the numbers (and percentages) of children in each group who had the rating of 'certainly applies' for each symptom
*Chi-square indicates significant difference between groups, $p < 0.05$

The overlap between the high-scorers on the subscales was not large. Two hundred and thirty-seven boys scored above the cut-off for emotional disorder; 32 (13.5 per cent) also scored high in conduct disturbance. Two thousand two hundred and twenty-five scored below the cut-off for emotional disorder; 240 of them (10.8 per cent) scored high in conduct disturbance. This association between the two types of deviance was not statistically significant at the 0.05 level on chi-square testing.

The mixed group was compared with the ED for symptoms of emotional disorder (see Table 3.1); and with the CD group for symptoms of conduct disorder (see Table 3.2). (The mixed group was not compared with ED for conduct disorder, or with CD for emotional

Table 3.2. *Symptoms of conduct disturbance in contrasting groups of conduct disorder*

	Conduct disorder		Mixed emotional and conduct disorder	
	N = 240		N = 32	
	No. with symptom	%	No. with symptom	%
Teacher rating				
Destructive	42	(18)	3	(9)
Fighting, quarrelsome	118	(50)	15	(47)
Disobedient	94	(39)	16	(50)
Lies	57	(24)	9	(28)
Steals	28	(12)	5	(16)
Bullies	68	(28)	13	(40)
Truants	4	(2)	1	(3)
Parent rating				
Destructive	31	(13)	6	(19)
Fighting, quarrelsome	39	(16)	7	(22)
Disobedient	63	(27)	17	(53) *
Lies	42	(18)	8	(26)
Steals	9	(4)	2	(6)
Bullies	22	(9)	4	(13)
Truants	1	(-)	0	(-)

The table shows the numbers (and percentages) of children in each group who had the rating of 'certainly applies' for each symptom
*Chi-square indicates significant difference between groups, p < 0.05

Table 3.3. *Percentages of children rated as 'certainly applies' for symptoms of teacher B(2) and parent A(2) scales*

		Emotional disorder N=205	Mixed N=32	Conduct disorder N=240	
		% with symptom definitely present	% with symptom definitely present	% with symptom definitely present	
Hyperactivity					
Restless	B(2)	12.7	43.8	47.9	*
Fidgety	B(2)	9.8	37.5	37.7	*
Cannot settle	B(2)	10.2	34.4	29.2	*
Restless	A(2)	25.5	34.4	32.6	
Fidgety	A(2)	16.2	15.6	16.5	
Cannot settle	A(2)	15.6	15.6	18.5	
Attitude					
Irritable	B(2)	2.9	50.0	24.6	*
Resentful	B(2)	2.4	46.9	24.2	*
Irritable	A(2)	17.8	25.8	16.9	
Somatic					
Aches or pains	B(2)	7.3	15.6	2.1	*
Headaches	A(2)	2.5	3.1	3.4	
Stomach aches	A(2)	6.9	6.3	0.8	*
Eating difficulty	A(2)	8.0	0.0	3.8	*
Sleeping difficulty	A(2)	9.1	6.3	1.7	*
Social					
Not much liked	B(2)	2.0	29.0	13.5	*
Solitary	B(2)	12.7	15.6	6.3	
Not much liked	A(2)	2.4	3.2	3.0	
Solitary	A(2)	8.3	12.5	3.4	

*Chi-square indicates significant difference between groups, p<0.05

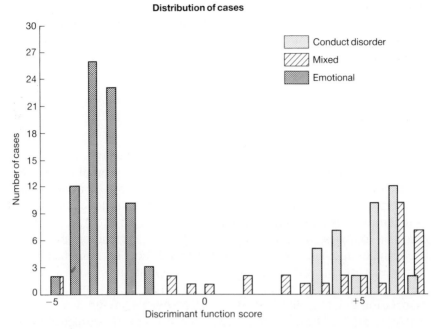

Fig. 3.1 Distribution of scores of boys, with conduct and/or emotional disorders, on a discriminant function separating conduct from emotional disorder.

disorder, as, of course, there were bound to be large differences because of the groups' definitions.) The purpose was to determine whether the symptoms of the mixed group suggested the blending of two independent dimensions or a different meaning of high scores. The symptoms of emotional disorder were indeed different when they occurred in combination with conduct disorder: depression of mood and fussiness were much more frequent in the mixed group, and the picture was the same for the independent ratings by teacher and parent.

By contrast, conduct disorder seemed to be much the same in form, whether or not emotional disorder was also present. The one possible difference was a trend to more disobedience towards parents when emotional symptoms were also present.

Comparison of the groups on the basis of other symptoms (that did not enter into the definition) is set out in Table 3.3. Symptoms of hyperactivity at school (but not at home) were more common when conduct disorder was also present; somatic symptoms were more common when emotional disorder was present; the mixed group had the characteristics of both. The mixed group was especially likely to be irritable and resentful, and to be disliked by other children.

The discriminant function analysis shown in Fig. 3.1 necessarily produced a sharp distinction between the pure groups. Mixed cases, for the most part, appeared to belong with conduct disorder on the basis of their symptomatology: 86 per cent were classified as CD, 14 per cent as ED.

DISCUSSION

The relative lack of overlap between children deviant on conduct disturbance and emotional disturbance subscales emphasizes the independence of the disorders. This study was able to avoid some artefacts that could inflate the association between types of deviance. Our sample was relatively homogeneous with regard to age, and was confined to males, and thus avoided an apparent correlation appearing between conduct and emotional disturbance simply because both were commoner in younger children. Since the population was epidemiologically defined it avoided the Berksonian effect of apparent association due to individuals with two problems having two reasons for referral. The boys who did show both types of problem would of course be expected to be referred to clinics with disproportionate frequency: our finding is therefore quite compatible with the experience of clinicians that there is a large group of mixed disorders in young children (Rutter & Gould, 1985; Wolff, 1985). The population studied was large enough to be able to take realistic cut-offs, giving numbers approximating to the prevalence of diagnosed psychiatric disorder in similar populations.

Although the two disorders are independent, there is still an issue of co-morbidity. The characteristics of the mixed cases did not conform simply to a blending of two independent dimensions. Most of them were not intermediate between the two pure groups but could be classified with the conduct-disorder group. Furthermore, emotional disorder was rather different in expression when uncomplicated than when combined with conduct problems.

The relationship between emotional and behavioural disturbance could well be causative. High levels of misery, irritability, resentment, and fussiness seem to be characteristic of the mixed group. This complex of feelings and attitudes would be very likely to motivate children to oppose adults and therefore to be rated as disobedient and conduct-disordered by adults. But this is not the only possible route, and developmental research into the mechanisms should be carried out. It is also possible that conduct disturbance creates adverse life circumstances and hence depression of mood; even though longitudinal studies (such as that by Schachar and co-workers (1981)) have not suggested that this is an important route. The mixed group is therefore likely to repay aetiological and longitudinal study—especially the fraction showing depression of mood.

The implication of these findings for dimensional schemes of classification, or for multiple-category schemes such as DSM–III-R, is the need to recognize how one dimension may be altered when there is deviance on another. The implication for single-category schemes such as ICD-10 is the desirability of creating a separate group of mixed cases for further study. For the remainder of the present study, emotionally disordered boys were excluded from all the groups to be studied intensively.

Application of detailed measures to selected behavioural groups

4 Developmental and cognitive contrasts

Previous chapters have shown the need to compare hyperactivity (HA) with the defiant, aggressive conduct disorders (CD). Although the two overlap, they are statistically separate and it is possible to compare them. Previous investigators, using smaller populations, have often resorted to partial comparisons, e.g. of hyperactive conduct-disordered children with the conduct-disordered who are less hyperactive (Sandberg *et al.*, 1978, Schachar *et al.*, 1981) or of hyperactivity with and without conduct disorder against anxiety disorder (Reeves *et al.*, 1987). The problem here is that children with mixed hyperactivity and conduct disorder show more severe hyperactivity than the hyperactive, and more severe conduct disorder than the conduct-disordered (see Chapter 2). A finding such as more social disadvantage, poorer cognition, and worse outcome in the HA-CD than the CD (Schachar *et al.*, 1981) could therefore characterize a severe degree of CD rather than HA in itself. The correct comparison is between HA and CD groups.

This chapter reports the differences between them on the measures of children's development. It compares them with mixed cases showing both problems and with controls without those problems. It also reports on a group of children with attention deficit who are not hyperactive, because this group should help to show whether attention deficit is the underlying problem in hyperactive children.

METHODS

Subjects

The sampling frame for this study is described in Chapter 2. Out of all the 6- and 7-year-old boys on the registers of schools in Newham, other than schools for those with severe learning disability, 2462 had teachers and parents who both completed the screening rating scales. Screening questionnaires were completed in the spring, so that teachers had had 6 months to get to know the children.

Four groups were selected for more detailed study: those with scores above cut-off for conduct disorder who also met criteria for pervasive hyperactivity (mixed); those who met criteria for pervasive hyperactivity but not conduct disorder (HA); those who met criteria for conduct disorder but not for hyperactivity (CD); and those who did not meet criteria (control). Subjects were excluded if they were at special schools for emotional or behavioural difficulties, or if they had scores on the emotional disorder subscale of either Rutter scale of 5 or greater (since these were considered to form a separate group and could have confounded the comparison of other groups, see Chapters 2 and 3). Forty-five subjects were randomly selected in each group from all the boys meeting the definition, in order to yield approximately 40 intensively studied subjects in each group.

In addition, all the boys who scored 1.5 or higher on the inattentiveness subscale of items from the Conners' classroom rating scale, yet scored 2 or less on both hyperactivity subscales, were selected as a fifth group of attention deficit without hyperactivity (AD) (N = 33).

These 213 boys were then traced through the schools, and their parents were visited and asked for consent to the remainder of the study. (Some other boys were also traced for other purposes, including an additional over-sample of purely hyperactive boys and some of those in special schools—see Chapter 7). The second stage of measures was then administered, at a mean of 9 months after the first (screening) stage (range 6–12 months). In all cases, the summer vacation had intervened between the questionnaire screen and the second stage; so the boys had new teachers, and were aged 6 years 6 months to 8 years 9 months.

The study was divided into two halves, half the borough being surveyed in one year and the other half in the following year. No differences were expected or found between the subjects in the two years and they were accordingly analysed together.

Second stage measures

Parental account of children's symptoms (PACS)

This standardized, investigator-based interview measure was developed as an instrument for the measurement of children's behaviour problems as seen at home (Taylor *et al.*, 1986*a*). It is administered by a trained interviewer. Parents are asked, not for their ratings of problems, but for detailed descriptions of what their child has done in specified situations over the previous week. Such situations are defined either by external events—e.g., watching TV, reading a book or comic, playing alone, playing with friends, going to bed, travelling—or by behaviours shown—e.g., crying, worried talk, tempers, fighting with siblings. The interviewers then make their own ratings, on the basis of their training and written definitions of the behaviours to be rated, on a 4-point scale (0 to 3) of severity and frequency in the previous week and previous year.

The subscales are:

(1) *Hyperactivity* This is made up of attention span (time spent on a single activity, rated separately for four different kinds of activity), restlessness (moving about during the same activities), fidgetiness (movements of parts of the body during the same activities), and activity level (rated for structured situations such as mealtimes and car journeys).

(2) *Defiance* This scale is composed of items concerning temper tantrums, lying, stealing, defiance, disobedience, truanting, and destructiveness.

(3) *Emotional disorder* This is made up from items of misery, worrying, fears, hypochondriasis, and somatic symptoms. It therefore relates to overt emotional stress, not to inferences concerning the emotional basis of symptoms.

Standardized teacher interview

An experienced child psychiatrist (S.S. or E.T.) administered a detailed, structured, standardized interview schedule to the child's regular class teacher; the same psychiatrist examined the child. Questioning covered a range of situations encountered in most such classes, including quiet group work (such as listening to a story), quiet and structured work on a set task, quiet work on a task of the child's choice, group work on set and voluntarily chosen tasks, and occasions involving the whole class. The teacher was asked for specific examples of behaviour, and the behaviour was then rated by the psychiatrist. Finally, at the end of the interview, the psychiatrist rated the presence or absence of all the behaviours specified in DSM–III as defining ADDH.

The interviewer of teacher, interviewer of parent, and the psychologist who tested the child were blind to each other's information and to group membership.

Psychological tests

A battery of psychological tests was given individually to children in their schools.

Intelligence Two subtests from the Verbal part of the Revised Wechsler Intelligence Scales for Children were given (similarities and vocabulary) and two from the Performance Scale (block design and object assembly).

Reading The Neale tests for accuracy and comprehension of reading short passages of text were given. Schonell tests were used, of single-word reading, when the Neale was too difficult. Each test is scored as a reading age. For accuracy and comprehension separately, two measures of impairment were derived:

(i) the delay in reading, i.e. the reading age subtracted from the chronological age

(ii) the discrepancy in reading, i.e. the reading age subtracted from the predicted reading age (predicted by a regression equation using chronological age and IQ).

Digit span The digit span subtest of the Wechsler scale was also given, the score being the maximum number of digits correctly repeated in order.

Continuous performance test This is designed to measure sustained attention on a task requiring vigilance (Erlenmeyer-Kimling & Cornblatt, 1978). The boys sat in front of a video screen on which a microcomputer program presented 440 representations of playing cards, each lasting 500 msec, with an interval of 1000 msecs between each presentation. The task was to detect when a picture was identical to the one immediately preceding it (as in the game of 'Snap'). This happened on 55 occasions during the test, randomly dispersed among the other stimuli. The subject was instructed to press a button when, and only when, the picture was identical. Errors were of two kinds: missing an identical pair ('omission error') and pressing the button when the two previous pictures were not identical ('commission error'). Two patterns of errors are calculated on the basis of signal detection theory (Pastore & Scheirer, 1974): observer sensitivity, which is low when the observer cannot distinguish 'signal' from 'noise'; and observer criterion, which is low when the observer sets a low threshold for responding.

Central–incidental learning test This test was developed from the theory and measures of Hagen and Hale (1973). The subject is asked to learn and recall test material ('central learning') and then, at the end of the test, to recall material from the test that did not form part of the instructions ('incidental learning'). The hypothesis is that selective attentional processes lead to better learning of the central test, at the expense of the incidental material; and that incidental learning is correspondingly better in people who lack the normal degree of selectiveness. In administration of the test, a board was shown to the child; on it were 16 pictures, 8 of animals and 8 of foods. A practice board was shown first to ensure that the task was understood. The child was instructed to learn the positions of one of the sets of pictures, either the animals or the foods. After a 15-second presentation, the board was covered and the child given 8 cards, each bearing one of the target pictures, to place in the correct location (the 'central' task). When this was done, he was asked similarly to locate cards with pictures of the 'incidental' stimuli. The score, for central and for incidental tasks, was the number of cards correctly placed.

The MFF-20 is a revised and more reliable version (Cairns & Cammock, 1978) of Kagan's 'Matching Familiar Figures' test of reflectiveness v. impulsiveness. A picture has to be matched to six slightly varying pictures, only one of which is identical with the target. There were 20 such presentations. The scores were the number of errors made in the test and the time taken between the presentation of the picture and the first response made, whether or not the response was correct.

The Paired Associate Learning Test reflects the ability to learn simple rote associations; it requires an efficient use of strategies in short-term memory. The procedure used was that of Swanson and Kinsbourne (1976): pictures of animals were presented from a slide projector, and for each animal the child was asked to remember the name of the zoo in which it lived. The list length was adjusted to the ability of the child by starting with a list of four animals and working up to the shortest list length that the child failed to learn completely after ten repetitions. The score was the percentage of incorrect responses during the learning of the final list. The 'percentage success rate' was the score subtracted from 100.

Summary scale of attentional performance

Several of the above measures were combined together into a composite scale. This was based upon the results of factor analysis of the measures, indicating a large factor upon which several tests loaded—the Continuous Performance Test observer sensitivity measure, the Paired Associate Learning Test percentage success rate, the central learning measure, the error rate in the MFF-20, and the digit span. These test scores were therefore converted to comparable values by logarithmic transformation followed by z-transformation. They were then summed to give the 'attentional performance' scale, on which a high score indicates a good performance.

Behaviour during testing

Direct observation of off-task behaviours during the continuous performance test was made according to the scheme reported by Sandberg and co-workers (1978). A continuous event-sampling record was kept of body movements and of gaze fixation away from the test in progress. Mechanical actometers were attached to the non-dominant arm and leg and the readings expressed per minute of elapsed time.

Physical examination

This included a scored developmental neurological examination (Yule & Taylor, 1987), assessment of clarity of speech, and scoring of minor physical anomalies in the structure of head, hands, and feet (Waldrop *et al.*, 1968). In the motor and speech tests a high score indicated a less competent performance.

Developmental history

This was taken from the parents and included a standardized, structured interview enquiry about the dates of reaching milestones in language development. Possible complications of pregnancy (hypertension, toxaemia, antepartum haemorrhage, physical illness), delivery (breech, forceps, prematurity, caesarean section) and the neonatal period (special care, seizures, difficulty in breathing) were asked for, and any noted were summed with unit weight to give a 'perinatal adversity index' for early development.

RESULTS

One hundred and ninety-four of the 213 boys (91.1 per cent) took part in the detailed investigation. The validity of the group definition was examined by comparing the group scores on the detailed behavioural measures taken in the second stage (Table 4.1). There were no significant age differences between groups.

In spite of the changes that had intervened since the screening, the groups were still different in their behaviour according to independent measures. The HA and HA-CD groups scored high on all measures of hyperactivity and inattention; CD and controls on none; AD only on measures of attention disturbance at school. Furthermore, there were no differences between groups in emotional symptomatology that might otherwise have confounded the comparison. The screening

Table 4.1. *Detailed behavioural measures in contrasting behavioural groups*

| Detailed behavioural measure | Group in initial survey | | | | | | | | | | |
| | Pure hyperactivity (HA) N=41 | | Mixed (HA-CD) N=38 | | Pure conduct disorder (CD) N=42 | | Attention deficit (AD) N=31 | | Control (N) N=42 | | |
	Mean	(S.D.)	Mean	(S.D.)	Mean	(S.D.)	Mean	(S.D.)	Mean	(S.D.)	
Classroom											
B(2) hyperactivity	2.8	(1.9)	3.7	(2.1)	1.8	(1.7)	1.9	(1.8)	1.5	(1.4)	*
TI hyperactivity	8.1	(5.4)	10.8	(6.6)	7.0	(5.7)	6.9	(6.1)	5.7	(4.3)	*
CRS attention disturbance	5.6	(3.3)	6.5	(3.4)	3.8	(3.2)	5.9	(2.9)	3.7	(3.1)	*
TI attention disturbance	13.8	(9.3)	16.1	(10.1)	10.1	(7.7)	12.0	(9.5)	8.2	(6.7)	*
B(2) conduct disturbance	1.3	(2.0)	2.8	(2.8)	2.2	(2.6)	0.7	(1.7)	0.5	(1.1)	*
TI conduct disturbance	18.3	(14.3)	32.7	(17.2)	22.6	(13.3)	13.8	(11.4)	12.8	(8.1)	*
B(2) emotional disturbance	1.4	(1.5)	1.4	(1.7)	0.9	(1.1)	2.1	(1.7)	1.5	(1.6)	NS
TI emotional disturbance	7.8	(3.1)	8.7	(2.6)	7.6	(2.7)	8.4	(2.7)	7.8	(2.9)	NS
Home											
A(2) hyperactivity	3.0	(1.9)	3.2	(1.6)	1.4	(1.3)	0.9	(1.0)	1.4	(1.5)	*
PACS attention disturbance	1.1	(1.0)	0.8	(0.7)	0.6	(0.6)	0.7	(0.7)	0.5	(0.5)	*
A(2) conduct disturbance	1.7	(1.7)	3.1	(2.0)	1.4	(1.3)	1.1	(1.3)	1.2	(1.4)	*
PACS conduct disturbance	0.8	(0.5)	1.1	(0.5)	0.8	(0.4)	0.8	(0.3)	0.8	(0.4)	*
A(2) emotional disturbance	1.3	(1.1)	2.0	(1.5)	1.4	(1.2)	2.0	(1.4)	1.6	(1.5)	NS
PACS emotional disturbance	0.5	(0.4)	0.6	(0.4)	0.5	(0.3)	0.6	(0.4)	0.5	(0.3)	NS

The table shows the mean and standard deviation on scales of behaviour from the B(2) (teachers) and A(2) (parents) rating scales, the CRS (Conners scale), TI (standardized teacher interview) and PACS (standardized parent interview), all administered in the second stage of the study.
*Analysis of variance shows significant difference between groups, $p < 0.05$

measures therefore gave a valid categorization of hyperactivity. The picture with regard to measures of conduct disorder was more complicated. The CD and HA-CD groups were still high on teachers' ratings of conduct disturbance, while HA, AD, and controls were not; so the distinction was valid according to the defining measure (even though the rater was different). However, parents' ratings and interviews in the second stage suggested that only the HA-CD were high on conduct disturbance; so the CD definition was only partly validated.

These measures of activity and attention did not include the behaviours in the DSM–III definition of ADDH that are taken to be measures of 'impulsiveness'. The reasons are given in Chapter 1, but the behaviours are still of interest in describing the psychopathology of children. We therefore enquired about them during the detailed interview with the teacher. Some of the items—'often acts before thinking', 'often ignores directions'—were very common in all the groups studied and did not show any significant differences between groups. The other items discriminated significantly between the controls and deviant groups, but were very similar in the HA and the CD. Thus, 'frequently calls out in class' definitely applied to 10 per cent of the controls, 29 per cent of the HA, and 26 per cent of the CD; 'often interrupts ongoing conversations' to 17 per cent of controls, 32 per cent of HA, 31 per cent of CD; 'difficulty waiting for his turn' to 2 per cent of controls, 15 per cent of HA, 26 per cent of CD; and an excessive need for supervision affected 29 per cent of controls, 56 per cent of HA, and 46 per cent of CD. The items are therefore not likely to be very helpful in the discrimination of different types of deviance, although they may obviously handicap educational progress.

The scores of each group of boys on the neurodevelopmental measures are shown in Table 4.2.

Minor complications of delivery had been quite common in all groups, but there was no indication of their being associated with behavioural status. Similarly, the index of perinatal adversity in the history, covering all recalled aspects of pregnancy and perinatal course, did not differentiate the groups. Delays in language development were frequently noted, but the differences were not significant. Nevertheless, the history of language delay was three times more frequent in the attention deficit group than in controls. The finding may be of interest, in that the detailed neurological examination of the children identified a current speech problem in more than half the attention deficit group; the difference from controls was statistically significant. This AD group was also different from controls in having a lower IQ. Their verbal IQ (mean 85) was lower than their performance IQ (mean 94), but

Table 4.2. *Neurodevelopmental and educational status of behaviourally defined groups*

	Pure hyperactivity (HA) N=41		Mixed (HA-CD) N=38		Pure conduct disorder (CD) N=42		Attention deficit (AD) N=31		Control (N) N=42		Pairwise comparisons with controls
	Mean	(S.D.)	Mean	(S.D.)	Mean	(S.D.)	Mean	(S.D.)	Mean	(S.D.)	
IQ	101	(14)	95	(12)	103	(17)	88	(11)	101	(13)	*AD<N
Perinatal adversity index	1.8	(2.2)	1.7	(2.4)	1.8	(2.0)	1.8	(2.4)	1.9	(2.3)	
Neurological examination	9.6	(8.1)	9.3	(7.4)	7.5	(6.3)	9.2	(7.9)	6.8	(5.5)	
Reading delay (in months)											
accuracy	−1.2	(13)	−7.7	(13)	0	(12)	−11	(10)	+2.4	(17)	*AD, HA-CD<N
comprehension	−3.3	(12)	−9.2	(12)	−2.5	(12)	−13	(9)	+0.6	(15)	*AD, HA-CD<N
	%		%		%		%		%		
Complications of delivery	37		41		46		29		24		
History of language delay	15		20		23		28		9		
Speech problem	26		29		22		52		12		**AD>N

The table shows the scores of the survey groups on IQ, historical measures of neurological adversity and language development, and current measures of motor co-ordination, speech, and reading.
*Analysis of variance significant between groups, p<0.05
**Chi-square significant between groups, p<0.05

Table 4.3. *Activity levels in behaviourally defined groups*

	Pure hyperactivity (HA) N=41		Mixed (HA-CD) N=38		Pure conduct disorder (CD) N=42		Attention deficit (AD) N=31		Control (N) N=42		Pairwise comparisons with controls
	Mean	(S.D.)	Mean	(S.D.)	Mean	(S.D.)	Mean	(S.D.)	Mean	(S.D.)	
Observed movements	2.2	(5.6)	2.5	(3.1)	0.6	(1.6)	1.1	(2.3)	0.7	(1.7)	*HA, HA-CD>N
Gaze off task	4.7	(8.4)	4.9	(5.6)	1.9	(3.8)	3.4	(4.2)	1.5	(2.2)	*HA, HA-CD>N
Wrist actometer	4.1	(3.3)	5.2	(6.7)	2.8	(1.9)	3.8	(6.8)	2.4	(1.6)	*HA-CD>N
Ankle actometer	7.6	(10.5)	7.7	(10.8)	4.3	(4.1)	3.8	(3.7)	2.7	(1.8)	*HA, HA-CD>N

The table shows the scores of the survey groups on laboratory measures of activity.
*Analysis of variance significant between groups, p<0.05

for both verbal and performance IQ the AD group was the only one scoring below controls. The attention-deficit group also differed in having poorer skills in the accuracy and comprehension of reading. While they were lagging behind in their achievement, this did not in fact differ significantly from that predicted by IQ, and the discrepancy between predicted and actual reading age was not significantly different between groups for accuracy ($F = 1.5$, df 4, 189) or comprehension ($F = 1.6$, df 4, 189). The mixed group (HA-CD) was also delayed below chronological age in the comprehension and accuracy of reading, but not below the achievement level predicted by IQ and age.

No other groups were significantly different from controls, and the hyperactive did not differ from the conduct-disordered.

Laboratory tests of activity and attention differentiated the groups (Tables 4.3 and 4.4).

Objective measures of activity level showed that the two hyperactive groups (HA and HA-CD) were the only ones to show significantly raised activity.

Most of our tests of abilities related to attention differentiated significantly between groups (Table 4.4). The attention-deficit group was most clearly distinguished, being significantly different from controls (and the conduct-disorder group) in digit span, paired associate learning, and central learning; and in addition those in this group were worse than those with conduct disorder on the matching familiar figures test. The mixed HA-CD group was impaired on digit span and was impulsive in the Matching Familiar Figures test, with a shorter response time and a higher number of errors than the non-hyperactive conduct-disordered group. The time taken to respond and the number of correct responses were significantly associated ($r = 0.65$, $p < 0.001$), and time was negatively related to the number of errors made ($r = -0.64$, $p < 0.001$).

The summary scale of attention performance discriminated significantly between the groups, even when the effects of general intellectual development were allowed for by covarying IQ. The attention-deficit and the mixed HA-CD groups were impaired by comparison with controls; and in addition the hyperactive group scored less well than the conduct-disordered. Since the impairment of the mixed group left some doubt about which of its components was linked to poor test performance, a correlational analysis was undertaken between dimensions of behaviour and the attention performance scale for the deviant boys of the HA, CD, and HA-CD groups only. Within this disruptive group, the Rutter subscale of classroom hyperactivity correlated significantly with lower levels of test performance ($r = -0.3$, $p < 0.001$), as did the Conners inattention scale ($r = -0.4$, $p < 0.001$),

Table 4.4. *Attention tests in behaviourally defined groups*

	Pure hyperactivity (HA) N=41		Mixed (HA-CD) N=38		Pure conduct disorder (CD) N=42		Attention deficit (AD) N=31		Control (N) N=42		Pairwise comparisons with controls
	Mean	(S.D.)	Mean	(S.D.)	Mean	(S.D.)	Mean	(S.D.)	Mean	(S.D.)	
CPT											
Sensitivity	0.9	(0.05)	0.9	(0.05)	0.9	(0.06)	0.8	(0.05)	0.9	(0.06)	
Threshold	0.8	(0.1)	0.8	(0.2)	0.7	(0.4)	0.8	(0.1)	0.8	(0.2)	
Digit span	9.0	(2.9)	7.4	(2.8)	9.5	(3.1)	7.2	(2.2)	9.4	(3.2)	*AD, HA-CD<N
PAL: % errors	40	(26)	39	(25)	33	(24)	46	(25)	30	(25)	*AD<N
MFF–20											
Time to first response (secs)	10	(5.4)	9	(4.7)	13	(8.2)	10	(5.9)	13	(8.7)	*HA-CD<N
No. of errors	33	(13)***	34	(12)	25	(10)	34	(12)	30	(12)	**
Central learning	4.6	(1.9)***	4.5	(2.0)	5.3	(1.9)	4.1	(2.2)	5.2	(2.0)	*AD<N
Incidental learning	3.0	(1.9)	2.8	(1.6)	2.3	(1.7)	2.5	(1.8)	3.0	(2.0)	
Attention performance	1.5	(21)***	–1.0	(20)	15	(21)	–2.0	(21)	11	(25)	**AD, HA-CD<N

Note The table shows the scores of the survey groups on CPT (continuous performance test), PAL (paired associate learning test), MFF-20 (revised matching familiar figures test), other tests of attention, and a combined scale of attention performance ranging from +30 to −30 with low or negative scores indicating below average performance.

*Analysis of variance significant between groups, p<0.05

**Analysis of covariance significant between groups after allowing for IQ, p<0.05

***HA significantly worse than CD

while the Rutter conduct disorder subscale was not significantly associated ($r = -0.09$). For parents' ratings of hyperactivity r was -0.3 ($p < 0.001$); while their ratings of conduct disorder correlated -0.1 (not significant). Similarly, the hyperactivity subscale from the PACS interview gave $r = -0.28$ ($p < 0.001$); conduct disorder $r = -0.05$ (not significant).

None of the measures from the continuous performance test distinguished between the groups. The overall probability of a correct detection was 0.7, of a false response 0.05; so the test difficulty had been set at an appropriate level and there was scope to find group differences. When the first 3 minutes of the test were compared with the last 9 minutes, the probability of hits and errors was the same in both parts for all children together ($F = 0.8$; df 1, 192); and the change between the two parts was not significantly different between groups ($F = 0.9$, df 4, 189). The lack of differences in performance was in spite of observed differences in behaviour during the test (Table 4.3).

The central-incidental learning test generated a better performance on the central task in all groups. Central learning differentiated the groups, but incidental learning did not; the ratio of central to incidental learning did not significantly differ between groups. There was a positive correlation between the central and incidental scores ($r = 0.32$, $p < 0.01$), and the correlation was similar in magnitude in all groups (between 0.18 and 0.39).

The presence or absence of psychiatric disorder in first degree family members was also compared between groups by chi-square comparisons, yielding no significant differences. A family history of behaviour problems in childhood, for example, was noted for 9 per cent of the HA, 13 per cent of the HA-CD, 11 per cent of the CD, 11 per cent of AD, and 13 per cent of controls. Histories of learning problems in childhood or of mental illness or antisocial traits in adult life did not differentiate between the groups.

DISCUSSION

The clearest neuropsychological differentiation was for the group of children with attention deficit who were not hyperactive. They were worse in both performance and verbal IQ, in reading ability, in speech development, in several tests of the attention battery, and in the composite attentional scale. Their cognitive impairment was general.

The differences between the hyperactive and conduct disordered groups were present but relatively subtle. The HA were more active during laboratory tests. This indicates that their hyperactivity is not only a matter of qualitative change, of bad behaviour, or of halo effects by raters; but represents a true and observable motor excess and an increase in task-incompatible actions. While this finding illuminates the nature of the activity change, it does not prove that the increased activity is a different type of disorder. The main distinction between the hyperactive and the conduct-disordered was the worse performance of the former on tests related to attention. It is likely that there is indeed a difference in the processes of learning in hyperactive children.

Neither of the pure groups was different from controls in neurological development, and there was no evidence for insults to the brain in previous development.

Only the mixed HA-CD, of all the groups of disruptive children, were different from controls; and this finding is compatible with other studies. McGee and co-workers (1984) found that only their mixed group showed impaired educational achievement. The likely reason, from the present study, is that the HA-CD is more severely affected in activity, attention, and conduct. Of these, it is the activity and attention change that generate the link with cognitive impairment. Milder levels of change in activity and attentive behaviour, as found in the HA group, are more weakly associated.

The findings did not support an association between these levels of hyperactivity and any sign of neurological abnormality. (Chapter 7 will show some associations when the behavioural groups are more rigorously defined.) The lack of a contributory history of brain injury in early life is in line with many other reports (Taylor, 1986). This, and the lack of motor abnormalities on examination, argue against an aetiology of minimal brain damage.

The nature of the attention deficit is not, on this evidence, a simple failure of the mechanism of sustaining focused attention. The length of the test was not a determinant of whether impairment would be shown, and errors were as likely at the beginning of tests as at the end. Indeed, it may be doubted whether the deficit can usefully be described as one of attention. There is certainly a tendency to make more errors in several tests. However, while central learning was worse, incidental learning was not better. One could not argue from this evidence that the deficit comes from an excessive breadth of attention, with less good performance on the intended task being associated with better processing of peripheral and task-irrelevant information. Whatever the difficulty is, it does not result directly from a shift of processing resources towards irrelevant stimuli at the expense of important ones.

Indeed, the striking aspect of inattentive behaviour was the range and severity of cognitive impairment with which it was associated. Chapter 8 will discuss possible mechanisms that could be involved. Future research should examine the course and determinants of the development of attentive behaviour through early childhood in order to understand the nature of pathological deviations.

5 Social background and family life

INTRODUCTION

Hyperactivity is usually conceived of as a biologically determined pattern of behaviour. Schachar (1986) has described the history of this belief, and demonstrated how much of it has been based upon the ideology of its proponents rather than upon objective scientific findings. Since aetiological research into diagnostic groupings of hyperactive children has shown very little that is specific to hyperactivity in its biological associations (Ferguson & Rapoport, 1983), one must ask whether the psychosocial environment in which children grow up might offer more practical clues to the development of this disorder.

There is a long tradition of studying social factors in hyperactivity, and some such studies have found positive results. As long ago as 1935, Childers examined hyperactivity as a symptom in behaviourally disturbed children; and concluded that psychological and social factors played a dominating role. Levin (1938) in an important but underrated paper described a large series of hyperactive children and concluded that, while organic factors were important at extreme levels of hyperactivity, lesser degrees were attributable to failures of socialization.

The succeeding half-century has seen a number of studies that reported associations with social factors (review by Taylor, 1986). Adoption, for example, was greatly over-represented in clinical series of hyperactive children from Canada and from California (Deutsch *et al.*, 1982); hyperactivity was more common in children who were known to have been unwanted because their mothers sought abortion during pregnancy (Matejccek *et al.*, 1985); and hyperactivity was disproportionately common in children who have grown up in institutions (Tizard & Hodges, 1978). Parents of children labelled as hyperactive were more negative about and disciplinarian towards their offspring (Hartsough & Lambert, 1982) and were less pressuring for academic development (Ackerman *et al.*, 1979). Experimental studies of mother–child interaction in the laboratory have often found differences when they work together on a task (Bee, 1967, Campbell, 1973). The differences, however, are not very consistent between studies, with mothers of hyperactive children appearing more directive in some and less directive in others. It is likely that the situation is a powerful determinant of the details of interaction; and this raises some doubt about whether the laboratory findings can be generalized to real life.

A more general doubt about most of the studies of children referred to clinics or special classes is that the controls have usually been taken from the general population. The referral process itself is determined by psychosocial factors (Shepherd, Oppenheim, & Mitchell, 1971). Most deviant children are not referred; and those that are referred are characterized by signs of stress in family life more than by extremes of deviance. It follows that statistically significant associations in this form of case-control study may result from the fact of the cases being referred, not from their being hyperactive.

We have previously reported findings in a correlational study of hyperactivity in children, all of whom had been referred to clinics because of disruptive behaviour (Taylor *et al.*, 1986*a*). Conduct disorder was correlated with signs of unsatisfactory family life; hyperactivity was not; but the difference could have resulted from referral bias. An epidemiological study suggested that hyperactivity was similar to conduct disorder in its associations (Sandberg *et al.*, 1980) but it was not possible to examine a hyperactive group without conduct disorder. Other studies of the general population of children have reported variously that teacher-rated hyperactivity is not linked to family size, status, or discipline (Campbell & Redfering, 1979) and is not linked to socio-economic status (Goyette *et al.*, 1978); or that it is more prevalent in disadvantaged areas (Glow, 1981). Pervasive hyperactivity was linked to low socio-economic status by the Isle of Wight survey (Schachar *et al.*, 1981) and by a Swedish community survey of 'minimal brain dysfunction' (Gillberg *et al.*, 1983). Hyperactivity defined by question-naire ratings was substantially more common in urban areas than rural in a Canadian study (Boyle *et al.*, 1987); and more common in deprived rural areas than in urban ones in China (Shen *et al.*, 1985).

All these studies have included conduct-disordered children among the hyperactive. This is a serious weakness for understanding the nature of hyperactive behaviour. The known, strong relationship between conduct disorder and family adversity (Wolkind & Rutter, 1985) may well have confounded all the results. McGee and co-workers (1984) were able to separate conduct-disordered and hyperactive groups; but they found no differences in socio-economic status or family relationships. It is possible that the relatively egalitarian nature of New Zealand society, from which McGee and co-workers reported, would make socio-economic status a less discriminating variable than it might be in more highly stratified societies.

In short, the existing scientific literature leaves it uncertain whether hyperactivity is associated with psychosocial factors, let alone whether there is any specificity to such an association, or whether there is a causal relationship, or whether any particular aspects of psychosocial adversity are more important than others.

We therefore report the results of the psychosocial measures that were administered for our community sample of boys at the beginning of their school careers. We have distinguished between measures of background social factors, which might be expected to affect all the children in a family; and measures of family relationships involving the child specifically under study.

METHODS

The design was the same as for the neurodevelopmental comparisons made in the previous chapter. Hyperactive and conduct-disordered boys were compared with each other and normal controls; and also with boys showing both problems and with a group showing inattentive behaviour solely. The measures of family life and relationships have been described elsewhere and have proved to have acceptable inter-rater reliability in clinical populations (Taylor *et al.*, 1986*a*). They were taken during the standardized interview with parents. *Parental coping* was assessed by questioning on actions taken in response to children's problem behaviours. The scale of coping is based, not upon the severity of the child's problem, nor on the interviewer's idea of what it is best for a parent to do, but rather upon the extent to which the parents' responses are planned and flexible in response to circumstances. It ranges from 1 (for highly efficient and adaptive coping skills) to 8 (for grossly inappropriate techniques). Ratings of 6 or more signified 'poor coping'. *Inconsistency* between parents was scored only when there was open argument about discipline and countermanding in front of the child. *Expressed warmth* and *expressed criticism* are based upon the parents' display of feelings towards their child during the interview itself and on counts of positive and negative remarks made. Scores at the most extreme level of warmth ('no warmth expressed') and criticism ('much criticism throughout interview') were taken to indicate deviance. *Ratings of the quality of parents' marriage* are made on an overall scale from 1 to 6 based upon systematic questioning on various aspects of the marital relationship, including warmth and criticism between parents, decision making, time spent together, communications, quarrels, and separations. A rating of 1 signifies a marriage typified by mutual concern and affection; a rating of 6 indicates either open antagonism or an absence of affection; ratings of 4 or more were taken to mark a 'poor marital relationship'. *Contact* between parent and child was assessed by the frequency of games, conversations, outings, and other joint activities. As expected, such activities were not frequent in this age group: a rate of less than once a week was taken as 'infrequent play'.

All these measures are based upon detailed and standardized schemes of interviewing, used in other investigations (Quinton, Rutter & Liddle, 1984). We established satisfactory, inter-rater reliability in advance, and continued reliability checks throughout the study. Mother's mental health was assessed by administering the Present State Examination Screening Schedule (9th edition). A family history of childhood problems and adult mental illness was taken after the parental interview from the main respondent.

Social background factors were assessed during the interview with the parents. Social class was defined by the Registrar-General's 1970 classification of occupations. Family size was defined by the number of children under 17 living in the household; 'overcrowding' was defined as more than 1.5 occupants per room. The parental situation referred to the people with whom the child was currently living, and a 'broken home' was defined as any situation other than living with two natural or adoptive parents. 'Previous separations' signified any spells of 1 month or longer when the child was living away from parents. 'Institutional care' was defined as present when the child had spent any period of his life in a children's home, regardless of the reasons for this and the legal status of the stay. 'Low occupational status' meant unskilled or semi-skilled manual occupation of the principal breadwinner.

Composite indices

Most types of adversity affected only a minority of children. However, it seemed probable that risk would increase with the number of stressors to which children were exposed. Accordingly, two additive indices were compiled.

A *social factors* index was calculated for each child by scoring one point for the presence of each of the following: overcrowding, unskilled or semi-skilled manual occupation of breadwinner, unemployment of breadwinner, unsatisfactory housing.

A *family relationships* index was calculated by scoring one point for each of the following: absence of warmth from mother, high expressed criticism from mother, infrequent contact with mother, infrequent play with father, high expressed criticism from father, interparental inconsistency.

Table 5.1. *Family relationships involving child*

	Pure hyperactivity (HA)		Mixed (HA-CD)		Pure conduct disorder (CD)		Attention deficit (AD)		Control	
	N=41		N=38		N=42		N=31		N=42	
	%	(N)	%	(N)	%	(N)	%	(N)	%	(N)
Mother										
Lack of warmth	33	(13/40)	37	(13/35)	18	(7/38)	24	(7/29)	20	(8/41)
High criticism	25	(10/40)	31+	(11/35)	8	(3/38)	7	(2/29)	12	(5/41) *
Poor coping	30+	(12/40)	23	(8/35)	13	(5/38)	21	(6/29)	* 5	(2/41) *
Low contact	30	(12/40)	34	(12/35)	35	(14/38)	28	(8/29)	34	(12/35)
Father/stepfather										
Lack of warmth	28	(9/32)	32	(9/28)	28	(9/32)	39	(9/23)	17	(5/30)
High criticism	22	(7/32)	32+	(9/28)	9	(3/32)	0-		3-	(1/30) *
Poor coping	18	(6/34)	25	(7/28)	19	(6/32)	9	(2/23)	13	(4/30)
Low contact	34	(11/32)	39	(11/28)	28	(9/32)	43	(10/23)	17	(5/30)
Between parents										
Inconsistency	18	(6/34)	26	(8/31)	19	(7/36)	4	(1/24)	15	(5/34)
Siblings										
Lack of play	21	(7/33)	15	(5/33)	8	(3/36)	12	(3/26)	8	(3/37)
Many fights	30	(10/33)	46+	(15/33)	19	(7/36)	35	(9/26)	24	(9/37)

The table shows the percentages of children in different groups showing the definite presence of adverse family factors. Numbers vary because of differences in family composition and informants interviewed.

*Differences between groups statistically significant by chi-square, p<0.05.

+ Group frequency significantly above that for all groups when tested by adjusted residuals, p<0.05

− Group frequency significantly below that for all groups when tested by adjusted residuals, p<0.05

RESULTS

Table 5.1 shows the frequency in each group of factors in family life that specifically involved the index child. Ratings for fathers should be read with caution: they were based upon less complete and systematic interviewing than were ratings for mothers, because of problems in persuading fathers to co-operate. Both for mothers and fathers, higher levels of critical expressed emotion were found for the mixed HA-CD group than for the others. Mothers' skills in coping with behaviour problems were worse in all the deviant groups than in controls, and particularly poor in the HA-CD. The HA-CD group were possibly characterized also by frequent quarrels with siblings.

There were few differences between the groups in background social factors (see Table 5.2). Housing was often of poor quality, but different aspects were troublesome to different families. Dissatisfaction with housing did not, however, relate to the behavioural status of the children; nor did the number of people per room or the availability of play space. This was in spite of the view, frequently expressed by parents to the interviewers, that unsatisfactory living conditions were the cause of children's problems. Many parents (but few children) were immigrants from overseas, the majority of the immigrants (68 per cent) being from the Indian subcontinent: however, the places of birth of parents and children did not predict the presence or nature of behavioural deviance (see section on 'Immigrant families', below). The only significantly associated factor was that the breadwinner's being in an unskilled or semiskilled manual occupation was more common in the group of inattentive but not hyperactive children (who were also of lower IQ as a group; see previous chapter).

Only one boy in the study had been adopted (in the HA group). The natural parents had divorced or separated for one quarter of the boys, but this bore no relation to their behavioural grouping.

The composite indices of social factors and family relationships are contrasted in Table 5.3.

Table 5.3 also shows other aspects of family life. The groups did not differ significantly in the frequency of a poor marital adjustment or the presence of depression in the mother defined by the Present State Examination. However, in many families there were lesser degrees of problems or symptomatology that fell short of caseness. Accordingly, the scale scores are shown. The HA-CD group mothers showed more depressive symptoms.

Since the HA-CD group was defined in part by mothers' ratings, its association with depression could have been the result of a rating bias. Depressed mothers might overestimate the hyperactivity and

Table 5.2. *Social background factors*

	Pure hyperactivity (HA)		Mixed (HA-CD)		Pure conduct disorder (CD)		Attention deficit (AD)		Control (N)	
	%	(N)	%	(N)	%	(N)	%	(N)	%	(N)
Overcrowding	13	(5/40)	9	(3/35)	13	(5/39)	10	(3/29)	12	(5/42)
Unsatisfactory housing	10	(4/40)	17	(6/35)	15	(6/39)	14	(4/29)	7	(3/42)
No play space	13	(5/40)	17	(6/35)	10	(4/39)	17	(5/29)	17	(7/42)
Large family	30	(12/40)	26	(9/35)	31	(12/39)	21	(6/29)	26	(11/42)
Low occupational status	37	(13/35)	31	(10/32)	50	(18/36)	63$^+$	(17/27)	28	(11/39) *
Unemployment	22	(8/36)	19	(6/32)	24	(8/34)	21	(5/24)	11	(4/35)
Immigrant mother	33	(13/40)	17	(6/35)	32	(12/38)	28	(8/29)	20	(8/41)
Separations from parents	18	(7/40)	14	(5/35)	13	(5/39)	10	(3/29)	10	(4/42)
Broken home	25	(10/40)	26	(9/35)	26	(10/39)	14	(4/29)	29	(12/42)

The table shows the percentages of children in different groups showing the definite presence of social factors, as defined in the text.
Numbers vary because of inapplicability of the factor or missing data.
*Differences between groups statistically significant by chi-square, $p < 0.05$.
$^+$ Group frequency significantly above that for all groups when tested by adjusted residuals, $p < 0.05$

Table 5.3. *Social and family factors in contrasting behavioural groups*

	Pure hyperactivity (HA)		Mixed (HA-CD)		Pure conduct disorder (CD)		Attention deficit (AD)		Control (N)		Pairwise comparisons
	N=40		N=34		N=40		N=29		N=42		
	Mean	(S.D.)	Mean	(S.D.)	Mean	(S.D.)	Mean	(S.D.)	Mean	(S.D.)	
Social factors index	1.1	(1.0)	1.0	(1.1)	1.2	(1.2)	1.3	(1.2)	1.1	(1.0)	
Family relationships index	1.7	(1.7)	2.1	(1.8)	1.3	(1.3)	1.1	(1.2)	0.9	(1.3)	*HA-CD>N,AD, CD, HA>N
Marital problems scale	2.1	(1.4)	2.4	(1.5)	2.5	(1.7)	2.2	(1.2)	2.1	(1.4)	
Maternal depression scale	2.3	(3.7)	4.1	(5.4)	2.0	(2.9)	1.1	(1.5)	2.7	(3.3)	* HA-CD>AD, HA,CD,N

*Differences between groups significant by analysis of variance, $p < 0.05$.
All significant pairwise differences, tested by Student-Newman-Keuls procedures, are shown.

conduct disorder of their children. Maternal and teacher ratings of behaviour were therefore examined separately for their associations with the maternal depression scale and the family relationships index. For mothers' and teachers' questionnaires separately, the scales of hyperactivity, conduct disorder, and emotional disorder were entered into a stepwise multiple regression with family relationships or depression as the dependent variable. (The choice of family relationships as a dependent variable here is open to question: it is not clear whether hyperactivity causes family problems or vice versa; but we are inclined to favour the former. In any case, the same conclusions are applied by partial correlation analyses). The results are summarized in Table 5.4. The associations between intrafamilial adversity and child behavioural deviance still held when the behaviour was independently assessed by the teacher. Hyperactivity was still associated with intrafamilial adversity, even after allowing for the degree of conduct disorder symptomatology: indeed, when conduct disorder and neuroticism were entered first into the equations, hyperactivity at home and school were still significant predictors of problems in family relationships.

The measures of various aspects of family life and relationships were themselves associated. The number of depressive symptoms in the mother correlated weakly but significantly with the family relationships index ($r = 0.17$, $p < 0.05$) and with the degree of marital problems ($r = 0.30$, $p < 0.01$). None of these measures, however, was significantly associated with the index of background social factors.

It was not possible to determine the direction of causality from these data. Discordant, angry family relationships might be caused by the child's own disruptive behaviours, or might be the cause of them, or both. Much of their potential importance is as factors that might determine the course of hyperactive children. Their association with short-term course of disorder is reported below (see 'Persistence of deviance in hyperactive children').

Immigrant families

There was a high concentration of immigrant families in the area studied. In nearly all of them, the child was born in the UK but the parents were born overseas. The largest group, 19.2 per cent of the families studied, came from the Indian subcontinent. A smaller number (13.3 per cent) came from the Caribbean, Africa, Ireland, or the Far East; but these latter were not sufficiently numerous to analyse separately.

Table 5.4. *Multiple regression of behaviour ratings against intrafamilial factors*

	Family relationships index		Maternal depression scale	
	T	Significance	T	Significance
Teacher ratings				
Hyperactivity	2.6	p<0.05	3.4	p<0.01
Conduct disorder	2.3	p<0.05	1.4	N.S.
Emotional disorder	1.0	N.S.	0.2	N.S.
Parent ratings				
Hyperactivity	3.6	p<0.01	2.5	p<0.05
Conduct disorder	3.4	p<0.01	3.1	p<0.01
Emotional disorder	2.0	N.S.	0.9	N.S.

The table shows Hotelling's T values and their significance for the association of hyperactivity, conduct disorder and emotional disorder (entered jointly in a stepwise regression) with an additive index of the number of adverse factors in family relationships, and, separately, with a scale of the number of depressive symptoms demonstrated by mother in the Present State Examination.

Table 5.5 shows the ratings of child behaviour for the different areas of origin. They are, in general, similar. The one discrepancy is that the immigrant children had, according to their parents, fewer emotional symptoms than the indigenous. Correspondingly, immigrant families were equally distributed across the behaviourally defined groups (pure hyperactivity, attention deficit, etc.). The differences between behavioural groups, which have been reported, are therefore not the result of ethnic differences.

Not surprisingly, there were differences between native British and immigrant families in family life and background. Table 5.6 compares the groups for the indices of psychosocial disadvantage and neurodevelopmental delay. The clearest finding is the lower rate of 'adverse' family relationships in the Asian and other immigrant groupings.

The global social adversity index conceals a more mixed pattern of psychosocial advantage and disadvantage in the Asian immigrants. Immigrant fathers were more likely to have further education after the age of 15 (57 per cent in the Indian families, 53 per cent in the other immigrants, 10 per cent in the British; $\chi^2 = 42.4$, df $= 2$, p<0.01); but there was a non-significant trend for them to be in lower status jobs (56 per cent in manual occupations in the Indian families, 39 per cent in the other immigrants, 36 per cent in the British;

Table 5.5. *Ratings of child behaviour in families from different areas of origin*

	Area of origin						
	UK		India		Other		
	N = 137		N = 39		N = 27		
	Mean	(S.D.)	Mean	(S.D.)	Mean	(S.D.)	
Teacher rating							
Hyperactivity	2.5	(1.9)	2.3	(1.5)	2.7	(1.9)	N.S.
Conduct disorder	2.3	(2.5)	2.4	(2.2)	3.1	(3.4)	N.S.
Emotional disorder	1.8	(2.0)	1.3	(1.5)	1.8	(2.4)	N.S.
Parent rating							
Hyperactivity	2.4	(1.9)	2.1	(2.0)	2.4	(2.1)	N.S.
Conduct disorder	1.0	(1.3)	0.9	(0.9)	1.1	(1.9)	N.S.
Emotional disorder	2.4	(1.6)	1.4	(1.4)	1.3	(1.3)	*

N.S.: no significant differences between groups
*$p < 0.01$ when differences between groups tested by one-way analysis of variance

Table 5.6. *Social and developmental factors in families from different areas of origin*

	Area of origin						
	UK		India		Other		
	N = 137		N = 39		N = 27		
	Mean	(S.D.)	Mean	(S.D.)	Mean	(S.D.)	
Social adversity index	0.9	(1.0)	1.3	(1.1)	1.1	(1.2)	N.S.
Family relationships index	2.6	(2.0)	1.5	(1.5)	1.8	(1.7)	*
Maternal depression score	3.0	(4.1)	1.8	(2.9)	2.0	(4.5)	N.S.
I.Q.	99.4	(15.4)	96.4	(13.3)	93.5	(17.6)	N.S.
Attention performance score	4.6	(21.2)	6.5	(25.9)	5.8	(25.0)	N.S.
Neurological score	10.8	(7.1)	7.1	(4.3)	7.6	(6.5)	*

*significant differences between groups ($p < 0.05$) tested by one-way analysis of variance

$\chi^2 = 4.4$, df = 2, p = 0.1) and they were no less likely to be unemployed (33 per cent of the Indian, 30 per cent of the other immigrants, 22 per cent of the British; $\chi^2 = 2.13$, df = 2, p = 0.3).

Family size was larger in the Indian group (mean 3.3 children under 17, S.D. 1.4) than in the other immigrants (mean 2.6, S.D. 1.4) or those born in Britain (mean 2.6, S.D. 1.2): F = 5.0, df = 2, 199; p < 0.01. In spite of this, overcrowding was equally common in all groups. Housing type and quality of housing were also similar between groups.

Family life was more stable for the Indian children: 95% were living with both natural parents, compared with 63 per cent in the other immigrant groups and 71 per cent in the British ($\chi^2 = 11$, df = 2, p < 0.01). A parent had had a previous marriage in only one of the Indian families (2.6 per cent), compared with 19 per cent of the other immigrants and 27 per cent of the British ($\chi^2 = 11$, df = 2, p < 0.01). The Indian children's mothers were less likely to have a psychiatric diagnosis. Fifteen per cent of the Indian mothers were identified as probable or definite current cases of affective disorder by the Present State Examination, 30 per cent of the other immigrants, 45 per cent of the British mothers ($\chi^2 = 12.4$, df = 2, p < 0.01).

There were no differences between the groups in the rated quality of the parents' marriage: the rates of severe marital problems were similar. However, marital discord was less common in the Indian (35 per cent) and other immigrants (27 per cent) than in the British (56 per cent) ($\chi^2 = 9.2$, df = 2, p < 0.01). When there were marital problems in the immigrant groups, they were more likely to be typified by a lack of affection and contact than by overt quarrelling or fighting.

It is therefore not surprising that the index of family relationships showed the immigrant families to have fewer problems than those with British-born parents. The item giving rise to the greatest difference was that of expressed criticism, which was present throughout interview for 16 per cent of Indian mothers, 4 per cent of other immigrants, 24 per cent of British mothers ($\chi^2 = 6.9$, df = 2, p < 0.05). There were no significant differences in expressed warmth or efficiency of parental coping, but marked inconsistency between parents was a little less common—not significantly so—in immigrant parents (11 per cent in Indian, 4 per cent in other, 19 per cent in British families: $\chi^2 = 5.3$, df = 2, p = 0.07). The Indian parents tended to spend less time in playing with their children: play involving mother with child took place less often than once a week in 53 per cent of the Indian, 38 per cent of the other, and 23 per cent of the British ($\chi^2 = 13.4$, df = 2, p < 0.01).

The presence of an immigrant subgroup characterized by social disadvantage but with less hostile, more distant relationships between child and parents could have confounded the overall results.

For example, a true association between behaviour problems and disadvantage would be obscured by including a group where high social disadvantage was offset by protective relationships. The behavioural group comparisons, reported in Tables 5.1 to 5.3, were therefore repeated for the subgroup of children whose parents had been born in Britain.

The pattern of results was not affected. Adverse social background (especially low occupational status) was still associated only with the 'inattentive' group; while the HA-CD group was still characterized by poor parental coping and high levels of expressed criticism from parents.

Persistence of deviance in hyperactive children

The presence of pervasive hyperactivity was an influence upon the short-term development of the boys. The initial screening groups were compared with the scores on the same questionnaires that were taken 9 months later, at the time of the detailed investigation (see Chapter 4). For this purpose, all the children in the second stage were taken as subjects, including the boys in the 'pure hyperactivity' group who had been over-sampled in the second stage.

The presence of disruptive behaviour (HA, CD, or both) at follow-up was significantly associated with the original screening group. In the mixed HA-CD group, disorder at follow-up was present in 81 per

Table 5.7. *Ratings predicting the persistence of behaviour problems in hyperactive children*

	Transient HA		Persistent HA		
	N = 31		N = 54		
	Mean	(S.D.)	Mean	(S.D.)	
Ratings from first questionnaires					
Teacher questionnaire					
Hyperactivity	4.2	(1.2)	4.1	(1.2)	N.S.
Conduct disorder	1.8	(1.9)	3.4	(2.7)	$p < 0.01$
Emotional disorder	2.1	(2.1)	2.0	(2.3)	N.S.
Parent questionnaire					
Hyperactivity	3.8	(1.1)	4.3	(1.2)	N.S.
Conduct disorder	0.7	(1.9)	2.0	(1.8)	$p < 0.001$
Emotional disorder	2.0	(1.5)	2.5	(1.9)	N.S.

cent (29/36); in the pure HA, 51 per cent (25/49); in the pure CD, 28 per cent (11/39); in the controls, 16 per cent (6/37) ($\chi^2 = 36.0$, df = 3, p < 0.001). The type of succeeding disorder was different in the HA and CD groups: to be in the CD group initially was only a risk factor for CD; but to be in the HA group carried risk for CD, HA and HA-CD later. Hyperactivity leads to conduct disorder; not vice versa.

Among the pervasively hyperactive boys who were followed up successfully (N = 85), there were 54 (63.5 per cent) who still met criteria for disorder at the second point of study. These boys, with persisting disorder, were compared with the 31 whose hyperactivity rating had been transient and did not show disorder at follow-up. Out of the scales of hyperactivity, conduct disorder, and emotional disorder from the first questionnaires, only conduct disorder predicted persistence of problems (see Table 5.7).

Transient and persistent hyperactivity could also be compared according to the psychosocial and neurodevelopmental measures taken at the second stage of detailed measures. These measures were not, of course, truly predictive since they were taken at the end of the 9-month period. They should be seen as possible associated factors, that could be the result of disorder persisting as well as the cause. Comparisons are set out in Table 5.8. Only intrafamilial problems were significantly associated with persistence of disorder. The separate factors of family life are set out in Table 5.9. The persistence of disorder

Table 5.8. *Factors associated with the persistence of behaviour problems in hyperactive children*

	Transient HA		Persistent HA		
	N = 31		N = 54		
	Mean	(S.D.)	Mean	(S.D.)	
Neurological exam	10.4	(7.9)	11.4	(6.9)	N.S.
IQ	97.8	(15.9)	97.7	(15.1)	N.S.
Attention performance	3.5	(23.7)	− 1.9	(17.3)	N.S.
Wrist movements	3.3	(1.8)	5.1	(6.1)	N.S.
Ankle movements	5.8	(9.5)	8.2	(10.9)	N.S.
Social factors index	0.8	(0.8)	1.0	(1.0)	N.S.
Family relationships index	1.8	(1.4)	3.2	(2.3)	*
Maternal depression scale	1.7	(3.5)	4.2	(5.0)	*

*Differences between groups statistically significant by analysis of variance, p < 0.05

Table 5.9. *Intrafamilial factors associated with the persistence of behaviour problems in hyperactive children*

	Transient HA		Persistent HA		
	N = 31		N = 54		
Intrafamilial factors	Number with problem		Number with problem		
Mental illness in mother (any level of severity)	7/31	(23%)	29/54	(54%)	*
Definite depression in mother (previous year)	3/28	(11%)	15/50	(30%)	N.S.
Either parent divorced previously	2/31	(6%)	16/34	(29%)	*
Lack of warmth from mother	5/30	(17%)	11/53	(21%)	N.S.
High criticism from mother	4/31	(13%)	24/54	(44%)	*
Poor coping by mother	6/31	(19%)	18/54	(33%)	N.S.
Poor coping by father	5/28	(18%)	10/48	(21%)	N.S.
Inconsistency between parents	4/24	(17%)	12/48	(25%)	N.S.
Poor marriage	2/30	(7%)	6/54	(11%)	N.S.

*Differences between groups statistically significant by chi-square, $p < 0.05$

in hyperactive children was not strongly associated with immigrant status of their parents, but outcome was a little better in immigrant families than in the indigenous British. Of 59 initially hyperactive children from British families, 42 (71.1 per cent) still had rated disorder at the second stage of testing; of 13 from the Indian subcontinent, 6 (46.1 per cent) still had disorder at follow-up; of 13 from other places, 6 (46.1 per cent) had persistent disorder. Differences between groups were not quite significant by chi-square testing ($\chi^2 = 4.9$, df = 2, $p = 0.08$).

The persistence of hyperactivity is therefore associated with conduct disorder symptoms and adverse family life: but these two factors are themselves associated.

When conduct disorder symptoms from the parent and teacher rating scales are both covaried, family life is still associated with persistence (F = 4.0, df = 1, 81; $p < 0.05$). Conversely, when family life is the covariate, persistence is still associated with teachers' ratings of conduct (F = 5.4, df = 1, 82; $p < 0.05$) and parents' ratings of conduct (F = 6.2, df = 1, 82; $p < 0.05$).

DISCUSSION

The main positive finding was the association of the mixed HA-CD group with signs of problems in family relationships. This was an association with hyperactivity as well as conduct disorder symptoms; and it was not due only to biased ratings by depressed mothers. The effect was on the persistence as well as the presence of disorder, and was in contrast with the lack of a relationship between the persistence of disorder and the presence of neurodevelopmental delays.

Since the findings were largely confined to relationships involving the child, it is possible that the behaviour problems gave rise to the high levels of negative expressed emotion and maternal depressive symptoms, rather than vice versa, and that the poor coping by parents reflected the problem that any parent would encounter in managing a high intensity of disruptive behaviour. It is hard to know which came first. Limited contemporary information from the health visitors' records was available about home circumstances in early childhood (see next chapter). The HA group was if anything less likely than other deviant groups to have had 'poor housecraft' noted in early childhood; yet both the HA and HA-CD groups had high rates of early behaviour problems. The evidence therefore does not support a common line of development in which disorganized family life is a sufficient cause of hyperactivity in children. Rather, there is evidence for an interactive process in which family factors help to determine the course of disorder once it has started.

The isolated finding that low socio-economic status characterized the inattentive group suggests a likely distinction between attention deficit and hyperactivity. The finding could have arisen from chance, given the number of comparisons made: but it would be in line with the possibility that social adversity creates a less stimulating environment and therefore a slowing of intellectual development. In the Isle of Wight studies using the same definition of hyperactivity, low social class was associated with pervasive hyperactivity: the finding is not replicated here. The difference between studies is unlikely to be due to smaller size or greater severity of the category of pervasive hyperactivity in the Isle of Wight (3 per cent v. the present 9 per cent), for the same lack of association holds for the still more restricted group of hyperkinetic disorder reported later (see Chapter 7).

The high rates of expressed criticism, poor parental coping, and maternal depressive symptoms are likely to be risk factors for the later development of the children. Indeed, even if they are initiated by a child's disturbed behaviour, they could still be a more powerful determinant of outcome than the degree of hyperactivity itself.

The short-term course studied here is compatible with this view, but does not definitely establish it. The associations could also be explained as a bad effect of persistent child problems upon family life. However, we have also found evidence for effects on course in a long-term study (Thorley, 1984): low socio-economic status predicted an antisocial outcome for both grossly over-active children and conduct-disordered controls in our fifteen-year follow-up study of clinic attenders. Family status was a more powerful predictor than was the presence of over-activity.

Longitudinal studies of groups differing in family relationships would clarify their significance as mediators of outcome. Controlled trials of modifying high negative expressed emotion and poor coping would be a useful means of studying preventive interventions.

6 Development and health of hyperactive children

This chapter describes three aspects of child health that could be related to hyperactivity. First, the records of community health screening are taken to give a contemporary picture of the earlier development of children in the survey. Second, a subgroup of hyperactive children and controls are compared to test some of the predictions of the theory that allergies cause hyperactivity. Third, the groups of children are compared to see whether hyperactivity causes accidental injuries.

PREVIOUS RECORDS OF HEALTH AND DEVELOPMENT

The previous chapters have indicated that attention deficit is often associated with histories of language delay, and hyperactivity with cognitive delays on examination, and alterations of family life. However, retrospective recall by parents is a weak way of eliciting information from the past. This chapter therefore draws on the contemporary records made of the health and development of the children in their first six years of life. The developmental records allowed us to address two questions: are birth injury and developmental delay associated with the development of hyperactivity? If so, do they contribute in the same way to hyperactivity and to other behaviour problems?

Method: Subjects

The same behavioural groups were studied as were compared in the two previous chapters. The parents of the selected cases were asked for permission for medical records to be examined. When consent was obtained, the developmental health record was examined by a child psychiatrist with experience in community health work and information coded blind to the group membership of the child.*

---

*This aspect of the survey was undertaken by Dr. M. A. Griffiths

Measures: Developmental records

The Community Health Department in Newham has an established programme of voluntary health and developmental screening for all pre-school and school-age children. This is undertaken by clinic-based teams of doctors and health visitors, and the health visitors also undertake routine home visits. Separate medical and health visitor records are kept together with maternity hospital records for each child, and medical or social services correspondence.

Most children received three developmental assessments during the pre-school period and, although there was some variation, the first check was usually at six weeks, the second at nine months, and the third at two-and-a-half years. A fourth developmental assessment was offered at school entry (age 5), but occasionally was not completed until the child was six years old. Each assessment included a medical and developmental examination, and a test of hearing and vision.

Information recorded

The investigator conducted a systematic review of developmental records of the study group, blind to subgroup membership of the child. The presence or absence of various obstetric factors was noted, as in Table 6.1, and the items were grouped to form two summary variables 'antenatal' and 'postnatal' complications. Quality of housecraft and

Table 6.1. *Obstetric variables derived from health records*

Antenatal complications	Postnatal complications
Number of miscarriages	Gestation more than 42 weeks
Family history of major disorder	Gestation less than 36 weeks
Parental consanguinity	Birth weight less than 1800 gm.
Maternal diabetes	Prolonged labour
Rubella in pregnancy	5 minutes delay in establishing
Hyperemesis requiring hospital care	respiration
Bleeding/threatened miscarriage	Floppy baby
Instrumental delivery	Prolonged poor sucking
Multiple birth	Convulsions, cyanotic attacks,
Psychiatric illness in pregnancy	hyperexcitability
	Hyperbilirubinaemia requiring
	treatment
	Major congenital abnormality
	Head injury/meningitis

quality of maternal care in the neo-natal period were coded from the health visitor record.

Four separate developmental examinations, spanning the ages from six weeks until the sixth birthday, were recorded. Language delay, co-ordination problems, poor hearing and vision, and growth retardation were noted at each examination as definite, possible, or absent. The four observations in each case were combined so that 'definite disorder of language' refers to the disorder being noted at any point from six weeks to six years. Only 'definite' abnormalities are analysed in the tables that follow.

Results

Two hundred and thirteen families from the second phase of the study were asked to consent, and permission to examine records was given by 194 (87 per cent). Of those 194 boys, 180 had had at least one developmental examination with adequately recorded clinical information by the age of 6 years, and 124 had been examined four times by this age. In the final analysis there were 41 boys in the conduct-disordered group: 38 boys in the hyperactive groups; 33 boys in the mixed conduct and hyperactive group; 30 in the inattentive group; and 38 in the control group.

Table 6.2 shows the numbers in each behavioural group with developmental abnormalities. Impairments in language development

Table 6.2. *Developmental and physical factors from health records*

	Hyperactivity (HA)		Mixed (HA-CD)		Conduct disorder (CD)		Attention deficit (AD)		Control (N)	
	N = 38		N = 33		N = 41		N = 30		N = 38	
	%	(N)	%	(N)	%	(N)	%	(N)	%	(N)
Language delay	34	(13)	39	(13)	24	(10)	60[+]	(18)	26	(10)*
Motor delay	11	(4)	15	(5)	5	(2)	13	(4)	8	(3)
Poor co-ordination	11	(4)	21	(7)	7	(3)	20	(6)	5	(2)
Hearing queried	29	(11)	21	(7)	27	(11)	40	(12)	26	(10)
Growth impaired	3	(1)	18[+]	(6)	5	(2)	13	(4)	3	(1)*
Vision queried	13	(5)	15	(5)	19	(8)	20	(6)	13	(5)

*Differences between groups statistically significant by chi-square, $p < 0.05$
[+]Group frequency significantly above that for all groups when tested by adjusted residuals, $p < 0.05$

Table 6.3. *Social and environmental factors from health records*

	Hyperactivity (HA)		Mixed (HA-CD)		Conduct disorder (CD)		Attention deficit (AD)		Control (N)	
	N = 38		N = 33		N = 41		N = 30		N = 38	
	%	(N)	%	(N)	%	(N)	%	(N)	%	(N)
Behaviour disorder	34	(13)	39[+]	(13)	10[−]	(4)	23	(7)	18	(7)*
Disorder in social development	29	(11)	33	(11)	22	(9)	47	(14)	16	(6)
Poor housecraft	3	(1)	15	(5)	10	(4)	17[+]	(5)	0[−]	*
Poor maternal care	8	(3)	9	(3)	0		3	(1)	3	(1)

*Differences between groups statistically significant by chi-square, $p < 0.05$
[+]Group frequency significantly above that for all groups when tested by adjusted residuals, $p < 0.05$
[−]Group frequency significantly below that for all groups when tested by adjusted residuals, $p < 0.05$

and physical growth, early behaviour problems, and poor housecraft were all unequally distributed over the behaviourally defined groups. Language development was impaired in the inattentive group, and there was a trend in the same direction for the two hyperactive groups.

When the language comparison was confined to those children whose parents had English as their mother tongue, or to those whose parents had been born in England, the difference between groups still had the same pattern and was still statistically significant.

There was no significant association between the antenatal and postnatal variables and the behaviourally defined groups (but see Chapter 7, as postnatal complications were more frequent in those with more rigorously defined hyperkinetic disorder).

The children's behavioural and social development and early family life are set out in Table 6.3. Poor housecraft was possibly associated with the inattentive group. Numbers, however, were too small for confidence in the statistical testing. Early onset of behaviour problems was significantly more common in the HA-CD group than the CD.

DISCUSSION

The results show that language delay is associated with the inattentive element, rather than the antisocial element of disruptive behaviour;

and that hyperactive behaviour problems are of earlier onset than is non-hyperactive, antisocial behaviour. Hyperactivity by itself is not necessarily associated with social disadvantage; but inattention (which in previous chapters was linked to cognitive impairment and low social class) probably is. These results confirm the conclusions of previous chapters from an independent data source, and extend the associations to an earlier stage of development.

Given that language delay and inattentiveness are associated, there are several possible mechanisms. A language delay might directly impair the normal development of attention, and increase a child's frustration; or inattentive behaviour might lead to language delay by discouraging parents from talking to their children and hindering the acquisition of language. Furthermore, both language delay and inattention may arise out of a common pathway of constitutional vulnerability or adversity in the psychological environment.

The uptake of developmental examination in the study group was very good. The families who consented to participate in the study were of course selected by this for their co-operativeness, and may have been similarly co-operative with earlier screening. Developmental information was missing only about those born overseas who came to the United Kingdom at school age; and was incomplete on those who moved about, both within and out of the borough, and those who did not attend clinics. Those families who move may have a disproportionate number of children at risk of later medical or educational handicap (Zinkin & Cox, 1976), and home-based health visitor assessments may be less predictive of later school-based problems than clinic assessments by doctors (Drillien & Drummond, 1983). Fortunately, the numbers about whom information was incomplete are small.

It ought to be useful to detect early language and behavioural problems in this way. Treatment could have been provided more widely. In its present form, screening for language and behaviour problems would identify only a minority of children who would later present hyperactive symptomatology. It would still be a useful screen, could yield a high-risk group for intervention, and (as will be seen in Chapter 7) would pick up a higher proportion of those who will develop the more serious hyperkinetic disorder.

HYPERACTIVITY AND ALLERGIES

The theory that hyperactivity is caused by the presence in the diet of artificial colourings, preservatives, and other harmful substances has a firm hold on many parents referring their children for help and has received much support from the mass media. In general, the scientific evidence has been weak and has indicated only that an occasional child may have an idiosyncratic, physically determined reaction to colourings; but that most of the children who seem to respond to elimination diets do so for psychological reasons (Taylor, 1986).

However, after the beginning of this study, a trial by Egger and his colleagues (1985) reported a dramatic response in a series of selected hyperactive children undergoing a blind trial of a radical diet excluding not only additives but a wide range of putative allergens. Most of the hyperactive children in their series had other symptoms of allergy. This raised the possibility that immunological processes played a part in the development of children's hyperactive behaviour. A recent controlled trial by Kaplan and co-workers (1989) has also found positive results when 24 hyperactive preschool boys were selected because of having physical signs and symptoms, and submitted to a diet eliminating a wide range of possibly harmful substances.

It was not, of course, feasible to mount a controlled trial of radical elimination diets in the setting of this epidemiological study. Physical symptoms of headache and stomach-ache were included in the population screen, and were very frequent complaints in the Egger and co-workers series. We therefore compared them across groups. They were commoner in emotional disorders than in other groups (Chapter 3), but their prevalence was not raised in conduct disorder or hyperactivity. However, this did not disprove the existence of other allergic disorders in the hyperactive.

We had used small subgroups of the study to pilot a questionnaire about symptoms of physical ill health in children. The questionnaire was administered to parents by a telephone interview. Some of the questions related to medical diagnoses of allergic disorder in the child or in relatives and to the symptoms of: red eyes, running nose, wheezing, night cough, mouth ulcers, abdominal bloating, colic, headaches, eczema, excessive thirst, and dark circles under the eyes. All of these are frequently cited by parents' groups as characterizing their supposedly hyperactive offspring.

We could therefore compare the two groups for whom the question-naire had been given. Twenty boys were randomly selected from all those who were pervasively hyperactive at both the screening stage and the second stage of enquiry. Twenty controls were randomly selected from all those boys who had scored below cut-off on both teachers' and parents' hyperactivity rating scales at both the screening and the second stage. Seventeen of the pervasively, persistently hyperactive group and 19 of the definitely non-hyperactive controls were successfully questioned.

The results did not suggest any excess of allergies in the behaviourally disturbed. Six out of the 17 pervasively and persistently hyperactive (PPH) had been diagnosed as allergic, compared with 4 out of the 19 controls. Five out of 17 PPH had been treated for allergy (usually with antihistamines) as compared with 4 out of 19 controls.

Out of 72 relatives for whom information was available in the PPH group, nine had a diagnosis of allergic disorder; compared with 7 out of 71 for the controls.

None of these differences was statistically significant when tested by chi-square or the Fisher exact probability method. Furthermore, with one exception, none of the symptoms enquired about differentiated the groups. The one exception was the symptom of 'excessive thirst'. This was mentioned for 7 out of 17 PPH and 1 out of 19 controls (Fisher exact probability $= 0.013$).

Though the groups were small, they were representative of their categories and gave an extreme of contrast between the presence and absence of hyperactive behaviour. The data therefore failed to support the claim that atopic symptoms were very much more common in hyperactive children. This does not exclude the possibility that allergic mechanisms played a part, but it does speak against one of the main arguments used to support the role of allergy. The trials by Egger and co-workers (1985) and Kaplan and co-workers (1989) were probably based upon atypical series of hyperactive children and should be interpreted accordingly.

ACCIDENTAL INJURIES

If hyperactive children are impulsive and reckless then they are likely to be at risk for accidental injury. Accidents are now the commonest cause of death in childhood, and victims are not a random sample of the population. Fatal accidents, like hyperactivity itself, are about three times as common in boys as in girls. If impulsive and over-active behaviour is indeed a risk factor then there would be possibilities for accident prevention. Modification of hyperactivity, or safety education directed towards hyperactive children, could be useful.

There is some contradiction in the existing studies. Langley (1984) reviewed them and concluded that the behaviour of children is at most a small factor in bringing about their accidents. On the other hand, a minor contribution to the risk would still be important if it could be removed. Furthermore, the evidence comes chiefly from epidemiological surveys of children with minor degrees of behavioural disturbance, and the risk might well be magnified in clinical disorder. A clinic study in London did find a high rate of accidents in hyperkinetic disorder by comparison with other conditions (Taylor *et al.*, 1986*b*).

We therefore asked parents, during the interview, about any injuries to the children in the previous year that had been serious enough to require medical attention. There was a difference between the behaviourally defined groups. Fourteen per cent of controls had had at least one such accident, compared with 31 per cent of the mixed HA-CD, 30 per cent of the HA, and 18 per cent of the CD ($\chi^2 = 9.4$, df = 4, p < 0.05).

Some of this difference could have been an artefact, since maternal report contributed to the behavioural definition as well as to the information concerning the accident. However, the teachers' ratings of behavioural deviance were clearly higher in those who had one or more accidents than in those who had not. The conduct disorder score from the Rutter teacher scale had a mean of 3.2 in the accident group, 2.2 in the others (F = 4.8; df = 1, 184; p < 0.05); the hyperactivity score was 2.9 against 2.1 (F = 7.4; df = 1, 184; p < 0.01). Independent sources of evidence therefore confirmed the association.

It is hard to be sure whether antisocial conduct disorder, or inattentive restlessness, or both, are responsible. A separate study* was therefore made, at the same time as the main study reported here, and has been presented elsewhere (Davidson & Taylor, 1987). All the cases of pervasive hyperactivity, and a normal control group, were taken as a cohort. At the end of a year we found out whether they had had any accidents by sending a follow-up questionnaire and by examining the records of all the casualty departments in the area. There was a significant rise in risk for those with conduct disorder as rated by parents (relative risk 1.79; 95 per cent confidence limits 1.20 to 2.67), but the relative risks as assessed by teacher ratings of conduct disorder (1.15) and hyperactivity (1.19) and parent ratings of hyperactivity (1.23) were not statistically significant.

It does seem that rule-breaking behaviour by children predicts a higher rate of accidents. Presumably the children with conduct disorder get a higher exposure to dangerous situations. We have shown that their behaviour is seen by their teachers as impulsive. This would make it harder to modify risk-taking behaviour directly. Reducing the hazards of the environment is still the most promising line for cutting the unacceptable death rate from accidental injuries.

---

*Dr. L. Davidson conducted this research

Redefinition of diagnostic groups with intensive behavioural measures

7 Hyperkinetic disorder: prevalence, definition, and associations

INTRODUCTION

Findings so far have been based upon groups defined by parent and teacher ratings. The differing associations of rated hyperactivity and conduct disorder suggest that both need to be recognized and included in diagnostic formulations. They do not, however, constitute valid ways of diagnosing individual children, and do not by themselves say what the detailed criteria for diagnosis should be. This chapter examines the category of boys who are identified individually as cases of hyperkinetic disorder.

The disorder, and some obstacles to studying it, have been described in Chapter 1. Its nature is uncertain: whether one should conceive of a rare or a common group of cases; one that is separate from conduct disorder or one in which conduct disorder (and perhaps other problems) are common; one that is confined to the neurologically disabled or one that is found in normal brains. In many respects its description is similar to that of the American Psychiatric Association's 1980 and 1987 descriptions of 'Attention Deficit Disorder with Hyperactivity' (ADDH) and 'Attention Deficit—Hyperactivity Disorder' (ADHD). ADDH is of course a well studied condition, at least in comparisons between referred cases and normal children. Nevertheless, existing studies have left it doubtful whether the diagnosis of ADDH has discriminative validity: direct comparisons with conduct-disordered children have suggested few differences (Koriath *et al.*, 1985, Reeves *et al.*, 1987). ADDH probably identifies a much larger group than does hyperkinetic syndrome (Taylor, 1986), but the diagnoses have not been directly compared.

The prevalence of hyperkinetic disorder is not known, and correspondingly the need for services is obscure. It is not clear whether more mildly hyperactive children are showing an attenuated version of the same disorder or a different sort of problem entirely.

We have therefore applied the measures, and criteria for caseness, that were developed for clinical populations, to an epidemiological study of boys at the beginning of their school careers. This chapter describes the developmental, psychosocial, and cognitive correlates of hyperkinetic disorder; its prevalence and its relation to other diagnostic definitions; and how it compares not only with normal children, but with those who have conduct disorder yet are not hyperactive.

METHODS

Subjects and design

After the screening of the population described in Chapter 2, contrast groups were selected for more intensive study (see Chapter 4). We are now concerned with the boys who were randomly selected from the four cells defined by the presence or absence of pervasive hyperactivity and by the presence or absence of conduct disorder (45 boys selected in each group); and with 20 pervasively hyperactive boys who were randomly over-sampled, in order to yield a group large enough for later follow-up. Of the resulting 200 cases, 183 (91.5 per cent) were traced, and agreed to participate; and complete information was available for 160 of them. The missing 23 represented cases where either the parent or the teacher could not complete the full interview or rating scale. The 40 cases with missing data were not significantly different from the 160 fully studied on any of the measures from the initial screening questionnaires.

Diagnostic criteria

Hyperkinetic disorder was defined, as in the clinical studies, as a score of 1.5 or greater on the hyperactivity-inattentiveness factor of Conners' Classroom Rating Scale, together with a score of 1.0 or greater on the hyperactivity scale of the PACS interview, and the definite presence of off-task activity during the psychologist's examination. (The third of these criteria proved to be redundant, as all cases meeting the CRS and PACS criteria also scored above cut-off on the observational scale.) The requirements for severity and pervasiveness across home and school observations were therefore met.

It should be noted that impulsive and rule-breaking behaviours are not a part of either scale. The questions are primarily about activity level and failure to sustain concentration on constructive activities.

The requirement for persistence over time was met by the additional criterion that cases should have been identified as pervasively hyperactive by the initial questionnaire screen. The requirement for absence of other psychiatric disorders was met by the exclusion of any boys scoring 4 or more on the emotional disorder subscale of either the parent or teacher rating-scale at either the first or second stage. No children presented autistic features, but these would also have been grounds for exclusion.

The absence of hyperkinesis (used for establishing a definitely non-hyperactive conduct-disorder group) was defined as a score of 0.5 or less on the PACS hyperactivity scale and 0.5 or less on the hyperactivity scale of the Classroom Rating Scale. This corresponded to scores within approximately 1 S.D. of the population mean. This group was also required to have scored 2 or less on both hyperactivity scales of the first stage. Many boys therefore fell into neither the 'hyperkinetic' nor the 'definitely non-hyperactive' group; they were classed as 'intermediate', and included with the definitely non-hyperactive for calculations of prevalence.

Conduct disorder was defined, following the manuals for DSM–III and ICD-10, as the presence of two or more of a list of undesirable behaviours. This was translated into the study measures as the definite presence of two or more items of antisocial behaviour, viz. frequent and marked tantrums, disobedience, aggressiveness, destructiveness, lying, stealing, disrupting other children's play, and defiance to adults—as described in the teacher and parent interviews.

Attention deficit disorder with hyperactivity (ADDH) was diagnosed from the numbers of symptoms of inattention, impulsiveness, and hyperactivity, as specified in the DSM–III manual (American Psychiatric Association 1980), recorded by the psychiatrist after the standardized teacher interview. The manual expressly requires information from the teacher to be the overriding source from which the diagnosis is made.

Analysis

The rate of hyperkinetic disorder was ascertained for the studied members of each screening group. The frequency of hyperkinetic disorder was then calculated, for each group, by multiplying the proportion, of those with hyperkinetic disorder to the number of the group that was studied, by the total number of that group in the whole population. The resulting frequencies were summed to give an estimate of the population prevalence of hyperkinesis (expressed for a population of 10 000).

The groups of hyperkinetic, conduct-disordered, and control children were then compared by analysis of variance. The dependent variables were the scores on the detailed measures of neurodevelopmental,

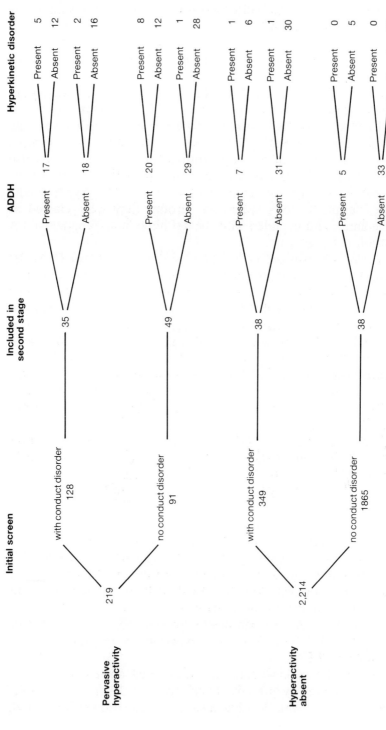

Fig. 7.1. Numbers of cases in screening categories of hyperactivity and conduct disorder and in diagnostic categories of attention deficit disorder with hyperactivity (ADDH) and hyperkinetic disorder.

cognitive, and psychosocial status. Pairwise comparisons of groups, using Student-Newman-Keuls procedures, were carried out only if analysis of variance showed a difference between groups.

Analysis of variance was also applied to examine the possibility of differences between varying definitions of hyperactivity. The first group comprised those who met criteria for pervasive hyperactivity but not for ADDH or hyperkinetic disorder; the second group was made up of those with ADDH who did not meet hyperkinetic disorder criteria; the third group was of hyperkinetic disorder. The final group was the children to whom none of these definitions applied.

RESULTS

Figure 7.1 shows the numbers of cases at successive levels of case identification. Hyperkinetic disorder was considerably less common than pervasive hyperactivity or ADDH. For the purpose of this comparison, the requirement of persistence over 6 months was omitted from the definition of hyperkinetic disorder: all but two of the cases had in fact been screened positive for pervasive hyperactivity. Those two subjects were omitted from other analyses, as they did not meet the criterion of persistence over time. The point-prevalence rate for hyperkinetic disorder was calculated, from its likelihood in the intensively studied groups and the frequency of those groups in the population, at about 172 per 10 000. The rate for ADDH was higher: about 404 per 10 000 if one considered only the children who had shown pervasive hyperactivity at the screening stage, or about 1660 per 10 000 if all groups were included. It will be seen that most children with ADDH in a population would not be detected by the questionnaires: but the screen was useful for the more restrictive hyperkinetic disorder.

A similar calculation gave a prevalence rate for conduct disorder of 740 per 10 000 who had shown conduct disorder at the screening stage, or 1850 per 10 000 if all groups were included. Forty-six per cent of those with ADDH, and 44 per cent of those with hyperkinetic disorder, met conduct disorder criteria as well.

Neurodevelopmental findings are set out in Table 7.1.

The hyperkinetic group proved to be different from the other two groups in neurological status. According to the examination, they were much clumsier, and many had an abnormality of speech. The history from parents also revealed more frequent delays of language in the earlier development of the hyperkinetic children. Their history was confirmed by the developmental screening records: there was at least one examination recorded before age 5 for 67 (88 per cent), and out

Table 7.1. *Neurodevelopmental status from examination, history and records*

	Conduct disorder (CD) N=25		Controls (N) N=35		Hyperkinetic disorder (HK) N=16		
	Mean	(S.D.)	Mean	(S.D.)	Mean	(S.D.)	
Neurological score	7.5	(5.6)	7.8	(6.2)	16.3	(6.6)	*HK>N,CD
Physical anomalies	4.1	(2.0)	4.2	(2.0)	5.2	(2.4)	
Obstetric suboptimality index	2.6	(2.6)	1.8	(2.0)	4.0	(3.6)	*HK>N,CD
	%		%		%		
Language delay							
on examination	4	(1/25)	6	(2/35)	31	(5/16)	**
parent's history	4	(1/25)	14	(3/35)	44	(7/16)	**
health records†	18	(4/22)	32	(10/31)	64	(9/14)	**
Poor coordination							
health records†	5	(1/22)	3	(1/31)	43	(6/14)	**
Perinatal problems							
parent's history	36	(9/25)	17	(6/35)	50	(8/16)	**
health records†	32	(6/19)	17	(5/30)	50	(6/12)	
Early behaviour problems							
health records†	27	(6/22)	10	(3/31)	71	(10/14)	**
Accidents							
2 or more in previous month	20	(5/25)	6	(2/35)	38	(6/16)	**

*Analysis of variance significant, p<0.05
**Chi-square significant between groups, p<0.05
†smaller N due to missing data in health records

of them a language delay was more often present in the hyperkinetic group; and more of the hyperkinetic had shown a problem of motor development.

The early origin of identifiable problems was also suggested by the history of adverse events in early life. The suboptimality index suggested a higher concentration of stress in pregnancy and birth. When individual items comprising the index were examined, the excess in the hyperkinetic group was all attributable to neonatal course: half of the group (8/16) had needed treatment in a special unit. The major types of neonatal problem were delays in establishing respiration, seizures, and 'jitteriness'. Minor physical anomalies were not significantly more frequent in any group; and if pregnancy factors were involved, they were not elicited by this study; so no prenatal factor was identified.

An immediately postnatal health record could be traced for 61 of the 76 children reported in Table 7.1, and this confirmed the pattern of perinatal problems reported by parents. Although the difference was not statistically significant with the reduced numbers ($p = 0.08$), the magnitude of difference between groups was very similar to parental recollection. Of the 17 known from records to have had perinatal problems needing special care, the parents of 16 had recalled such problems during the interview.

The physicians carrying out developmental checks had much more frequently noted behaviour problems in the first five years of life for the children who were later identified as hyperkinetic than for the controls, or even for the conduct disorder group. However, no differences had been noted between groups in physical illnesses (including deafness and allergic conditions).

Symptoms of physical illness were similar in all groups, according to parental report. However, accidental injuries were reported more commonly from the hyperkinetic group. These accidents were not severe: only five of all of the accidents had led to hospital treatment.

Cognitive status

The associations of the different groups with psychological test performance are set out in Table 7.2.

The results were clear-cut: most of the tests showed the hyperkinetic children to perform less well than the others. Where the differences were not significant, as in the continuous performance test, the trend was in the same direction. The conduct-disordered group was not significantly different from controls on any test. Pairwise comparisons showed that, for all the tests with significant analyses of variance, the

Table 7.2. *Psychological test results*

	Conduct disorder (CD)		Controls (N)		Hyperkinetic disorder (HK)		
	N = 25		N = 35		N = 16		
	Mean	(S.D.)	Mean	(S.D.)	Mean	(S.D.)	
IQ full scale	102	(14)	104	(13)	87	(17)	*
Verbal	98	(15)	99	(16)	85	(19)	*
Performance	106	(14)	108	(14)	92	(17)	*
Reading delay (months)							
Accuracy	+2.7	(15)	+1.0	(16)	−11	(11)	*
Comprehension	0	(11)	−1.7	(14)	−12	(11)	*
Continuous performance							
Observer sensitivity	0.91	(0.06)	0.91	(0.07)	0.89	(0.05)	
Observer criterion	0.72	(0.46)	0.73	(0.25)	0.76	(0.13)	
Digit span	9.2	(2.7)	9.6	(3.2)	7.2	(2.9)	*
Paired associate (% errors)	29.8	(27.9)	31.8	(24.0)	56.4	(22.8)	*
MFF-20							
Latency (secs)	13.5	(9.8)	12.1	(7.8)	7.2	(4.3)	*
Errors	27.3	(11.9)	28.2	(10.6)	38.7	(9.5)	*
Central learning	5.3	(2.3)	4.7	(2.0)	4.4	(1.4)	
Incidental learning	2.8	(2.0)	2.6	(1.7)	3.0	(2.1)	
Attention performance summary scale	+13	(25)	+11	(22)	−10	(15)	*
Activity							
Movements during testing	1.4	(3.6)	0.7	(1.6)	3.9	(7.8)	**
Actometer reading	3.0	(1.5)	2.3	(1.2)	4.9	(4.4)	**
Gazes away from test	3.4	(8.3)	1.9	(2.9)	6.7	(8.6)	**

*Analysis of variance significant, $p < 0.05$
**Analysis of covariance significant, < 0.05, with IQ as covariate

hyperkinetic children performed worse than the controls and the non-hyperkinetic conduct-disordered. The impairment extended to IQ measures; and, when IQ was allowed for, none of the tests of attention (but all the tests of activity) distinguished significantly between groups. On the continuous performance test, the similarity of performance by each group was in spite of marked differences in inattentive behaviour during that test.

The reduction in mean IQ seemed to reflect a small influence over the whole group rather than a small number of severely retarded individuals. The lowest measured IQ was 64, and only 3 children (2 in the hyperkinetic group and 1 in the conduct-disordered) had IQs less than 70.

Psychosocial factors

Factors from the family's social background did not differentiate the groups. The occupational status of the principal breadwinner, his or her unemployment, the parents' education, whether or not the parents were born in the UK and whether or not English was their first language, were all similar across disorders and controls. Housing was often unsatisfactory. Seven of the 76 families (9.2 per cent) were overcrowded (living more than 1.5 people to a room), and for 10 (13.1 per cent) the housing was in poor repair. However, there was no large or statistically significant difference between groups in people per room, type of housing, quality of housing, access to play space, or dissatisfaction with housing. Broken homes, remarried parents, previous separations, and institutional care, all occurred with similar frequency in all groups.

Psychological disorders in family members were found in all groups. Only mothers were examined directly, by means of the Present State Examination; and depression was the only diagnosis encountered in more than three mothers. 'Definite' depression in the previous year (corresponding to the level seen in psychiatric cases) was more common in both disorders; it had affected 4 (29 per cent) of the mothers of hyperkinetic boys, 6 (29 per cent) of those with conduct-disordered boys, and 1 (4 per cent) of control mothers ($\chi^2 = 4.7$, df $= 2$, n.s.). Fathers were asked more briefly about previous behavioural or mental health problems; there was no difference between groups.

Information about all first degree biological relatives (mothers, fathers, and full siblings) was taken from the respondent at the PACS interview, and set out in Table 7.3. The conduct-disordered group was the only one to show an excess of affected relatives, confined to conduct problems (i.e. definite patterns of persistently antisocial conduct in childhood or delinquency in adolescence).

Measures of family life and relationships at the time of the interview showed a number of differences between groups. The rates of various types of adverse family factors are set out in Table 7.4.

The hyperkinetic boys' mothers expressed more hostility towards them and coped less well with their behaviour problems; and these boys quarrelled more often with their siblings. For fathers it was only

Table 7.3. *Number of affected first-degree relatives*

	Conduct disorder (CD)		Controls (N)		Hyperkinetic disorder (HK)		
No. of probands	25		35		16		
Mean number of 1° relatives with information	3.3		3.2		3.0		
Relatives showing	Mean	(S.D.)	Mean	(S.D.)	Mean	(S.D.)	
Problems in learning	0.08	(0.28)	0.06	(0.24)	0.19	(0.40)	
Conduct problems	0.36	(0.70)	0.09	(0.28)	0.06	(0.25)	*CD>N,HK
Substance abuse	0.04	(0.20)	0.09	(0.28)	0.06	(0.25)	
Mental illness	0.04	(0.20)	0.03	(0.17)	0.13	(0.34)	

*Difference between groups significant by analysis of variance, p<0.05

Table 7.4. *Current family relationship measures*

	Conduct disorder		Controls		Hyperkinetic disorder	
	N = 25		N = 35		N = 16	
	%	(N)	%	(N)	%	(N)
Ratings of mother (N = 74)						
Lack of expressed warmth	24	(6/25)	18	(6/34)	40	(6/15)
High expressed criticism	20	(5/25)	6	(2/34)	33	(5/15) *
Poor coping	28	(7/25)	15	(5/34)	44	(7/15) *
Infrequent play	28	(7/25)	32	(11/34)	33	(5/15)
Rating of father (N = 64)						
Infrequent play	10	(2/20)	25	(8/31)	31	(4/13)
Sibling ratings (N = 67)						
No positive interactions	9	(2/23)	10	(3/30)	36	(5/14) *
Daily serious quarrels	29	(7/23)	7	(2/30)	50	(7/14) *
Interparental ratings (N = 64)						
Disagreement about discipline	30	(6/20)	3	(1/31)	23	(3/13) *
Poor marital relationship	35	(7/20)	9	(3/31)	15	(2/13) *

Note Where there were missing data on a measure (e.g. due to absence of a father), the percentages given are of those with non-missing data and their total number is given in parentheses for each measure.
*Differences between groups significant by chi-square, $p < 0.05$

possible to rely on the information about the quantity and type of contact between them and their sons, which did not vary significantly between groups: interviews with them were seldom sufficient to make satisfactory ratings of expressed emotion and coping skills. The conduct-disordered group was intermediate on expressed emotion and coping, but both it and the hyperkinetic group showed high levels of marked disagreement between parents over discipline. The conduct-disordered boys specifically were more likely to come from families where the marital relationship was poor.

Social adjustment

The categories of disorder had been defined by behavioural patterns, not by social maladjustment. It was therefore necessary to ask whether anybody considered the children to have any problems, or whether the 'disorders' we had defined were only personality variants. On the whole, and in spite of the quite high levels of hyperactive and antisocial symptomatology, the parents did not perceive the home behaviour of the children as a 'problem'. Hyperactivity was seen as a 'serious' problem for just 2 (12.5 per cent) of the hyperkinetic, 2 (8 per cent) of non-hyperactive conduct disorder, and none of the control group; it was seen as a 'minor' problem for 8 (50 per cent) of the hyperkinetic, 1 (4 per cent) of conduct disorder, and 2 (7 per cent) of controls; the remainder had 'no problem' ($\chi^2 = 21.1$, df $= 4$, p < 0.01). Conduct disorder was seen as a 'serious' problem for 3 (19 per cent) of hyperkinetic, 6 (24 per cent) of conduct disorder and none of the controls; and as a 'minor' problem for 8 (50 per cent) of hyperkinetic, 5 (20 per cent) of conduct disorder, 11 (32 per cent) of control ($\chi^2 = 13.1$, df $= 4$, p < 0.01). Only 4 parents in the hyperkinetic group (25 per cent) and 2 in the conduct-disordered, (8 per cent) had sought, or considered seeking, advice from the health services.

By contrast, parents did have considerable concern about their children's progress at school, seeing serious learning problems in 9 (56 per cent) of the hyperkinetic, 5 (20 per cent) of the conduct-disordered and 2 (6 per cent) of controls, and serious behaviour problems in 8 (50 per cent) of the hyperkinetic, 3 (12 per cent) of the conduct-disordered and 1 (3 per cent) of controls.

The concern was shared by teachers, who saw a marked educational problem for 15 of the 16 hyperkinetic, 9 of the 25 conduct-disordered and 1 of the 35 controls. The teachers also saw differences between the groups in their social behaviour with peers: they described overall problems of mixing with other children as severe in 6 of the 16 hyperkinetic (37.5 per cent), 5 of the 25 conduct-disordered children (20 per cent), and only 1 of the 35 controls (3 per cent) ($\chi^2 = 10.4$, df $= 2$, p < 0.01). However, the hyperkinetic children were not isolated or uninterested in their peers: only 2 of them (and 2 conduct-disordered children) were described as not interacting with other children. Rather, 6 of them (37.5 per cent) were seen as friendly but having no special friends (by comparison with 4 per cent of conduct-disordered and 20 per cent of control groups) whereas 50 per cent had one or more friends (compared with 88 per cent of the conduct-disordered and 78 per cent of controls) ($\chi^2 = 16$, df $= 2$, p < 0.01). The hyperkinetic were more often described as having severe problems in co-operating with others in play (hyperkinetic 64 per cent, conduct-disordered 18 per cent, controls 18 per cent, $\chi^2 = 14.1$, df $= 2$, p < 0.01); and in following the rules of games (hyperkinetic 50 per cent, conduct-disordered 8 per cent, controls 6 per cent, $\chi^2 = 17.9$, df $= 2$, p < 0.01). Correspondingly, they were often the victims of teasing by other children rather than instigators. Fifty-six per cent of the hyperkinetic children were described as being teased more than the other children, compared with 20 per cent of the conduct-disordered and 11 per cent of the controls ($\chi^2 = 12.5$, df $= 2$, p<0.01).

Subdivision of hyperkinetic disorder

An attempt was made to compare two groups of hyperkinetic children, with and without conduct disorder symptoms. The numbers, however, were too small for the comparison to have much power. Furthermore, the comparison was somewhat artificial. Although it was possible to define a group (9 out of 16) below cut-off for conduct disorder, even that group scored above the control group on the teachers' rating scales and interview measure of conduct disorder. It was therefore likely that the comparison was between greater and lesser degrees of conduct disorder, and that some degree of oppositional, disruptive behaviour in the classroom was a usual concomitant of hyperkinesis. The means of the more conduct-disordered hyperkinetic cases were close to that of the less conduct-disordered for the neurological examination (17.9 v. 15.0), IQ (88 v. 87), attention performance scale (-7 v. -12), social adversity index (1.5 v. 1.3), and number of rated peer problems (3.4 v. 4.0). None of the differences between groups was statistically significant when analysed by Student's t test. In short, the non-conduct-disordered hyperkinetic were more impaired neurodevelopmentally and socially than were the non-hyperkinetic conduct-disordered. It is therefore not the case that greater severity of conduct disorder was the reason for the differences between the hyperkinetic and the conduct-disordered children.

Different definitions of hyperkinesis

The hyperkinetic disorder considered so far in the analysis is less common and more severe than some of the categories used by clinicians and researchers.

Accordingly, Table 7.5 sets out the comparison between groups defined by the second stage of detailed behavioural measures. For the measures from the neurological examination and history, the findings are clear. The children with hyperkinetic disorder show more abnormalities than all the other groups, and the other groups statistically do not differ significantly, from one another. The same applies to IQ. By contrast, peer problems are progressively more frequent at higher levels of definition. For the composite scale of attention performance, the ADDH group is intermediate. It is less impaired than the hyperkinetic but more impaired than the controls.

Table 7.5. *Comparison of groups at different levels of definition of hyperactivity*

	I. Non-hyperactive (N) N=42		II. Pervasive hyperactivity (H) N=44		III. ADDH N=35		IV. Hyperkinetic disorder (HK) N=16		Group comparisons
	Mean	(S.D.)	Mean	(S.D.)	Mean	(S.D.)	Mean	(S.D.)	
Neurological examination	6.8	(5.0)	10	(6.4)	9.7	(7.2)	16	(6.6)	* HK>N,H,ADDH
Suboptimality index	1.7	(2.2)	2.3	(2.5)	2.3	(2.0)	4.0	(3.6)	* HK>N,H,ADDH
IQ	104	(14)	101	(13)	99	(17)	87	(17)	* HK<N,H,ADDH
Attention performance	+14	(23)	+5	(20)	+1	(19)	−10	(15)	* HK<N,H, ADDH<N
Peer problems	1.6	(1.9)	2.2	(2.0)	3.0	(2.0)	3.8	(2.0)	* HK<N,H, ADDH>N
Social adversity index	1.0	(0.9)	1.4	(1.0)	1.2	(0.9)	1.5	(1.0)	* HK>N,H, ADDH>N
	%	(N)	%	(N)	%	(N)	%	(N)	Adjusted Residuals
Perinatal problems									
History	20	(8/41)	32	(14/44)	20	(7/35)	50	(8/16) **	HK
Records	23	(8/35)	32	(10/31)	36	(10/28)	50	(6/12)	
Language delay									
Examination	5	(2/42)	2	(1/44)	3	(1/35)	31	(5/16) **	HK
History	12	(5/41)	14	(6/44)	23	(8/35)	44	(7/16) **	HK
Records	27	(10/37)	29	(10/35)	26	(8/31)	64	(9/14) **	HK
Early behaviour problems									
Records	13	(5/37)	26	(9/35)	40	(12/30)	71	(10/14) **	N,HK

Note Cases are excluded from each group if they meet criteria for a higher-level group, so each case appears once only in the Table.
*Difference between groups significant by analysis of variance, $p < 0.05$. All significant pairwise comparisons are shown.
**Difference between groups significant by chi-square, $p < 0.05$. All groups with significant adjusted residuals are shown

Since the impairment of the ADDH group was not accompanied by a significant reduction of IQ, the possibility arose of a more subtle and specific disorder of cognition in this group. Accordingly, those cases who met criteria for ADDH but not for hyperkinetic disorder were compared with non-hyperactive controls on the individual test scores from the psychometric test battery as listed in Table 7.2. The ADDH performed worse than the non-hyperactive on digit span (7.8 v. 9.7, t = 3.0, df 74, p < 0.05), errors on the MFF-20 (34 v. 26, t = 3.4, df 74, p < 0.05), and latency on the MFF-20 (10 v. 14 seconds, t = 2.2, df 74, p < 0.05). There were no significant differences in the observer sensitivity or threshold, the measures from the central-incidental learning test, or the paired associate learning test. The attention performance scale, the digit span, and latency and errors on the MFF-20 all showed significant differences, between ADDH and controls, by analysis of covariance when IQ was covaried. It therefore appeared that ADDH was associated with a pattern of cognitive impairment that was similar in kind to that of the hyperkinetic, but less in degree, and not associated with impairment severe enough to be manifest in IQ tests.

DISCUSSION

Nature and frequency of hyperkinetic disorder

Our findings are compatible with our suggestion from clinical research that hyperkinesis is a different sort of disorder from conduct disorder and is much more closely linked to neuropsychological dysfunction. It has an early onset, as contemporary health records of behaviour problems bear witness, and its association with perinatal disorder suggests that its causes lie early in development. Affected children tend to show motor clumsiness and cognitive impairment; their early development of language is delayed; their high activity levels can be measured directly in structured settings as well as being evident in qualitative ratings. These findings do not apply to non-hyperactive conduct disorder; and indeed most of them do not apply to the lower levels of hyperactivity defined by the categories of 'ADDH' or 'pervasive hyperactivity'.

Accordingly, a high threshold for the diagnosis is appropriate for a researcher who wishes to define a neuropsychiatric disorder for biological investigation; and for a clinician who intends the diagnosis to imply a form of neuropsychological dysfunction. The criteria adopted in this study, and partially validated by the predictive value of the resulting group, are close to the spirit of the draft ICD-10 criteria (WHO, 1988).

Hyperkinetic disorder in this sense is not a frequent condition. Indeed, the prevalence estimate of 172 per 10 000 is based upon a population likely to have a particularly high rate, and is therefore an overestimate of a generally applicable figure. To begin with, it is based upon males only. Several studies have suggested a 4 : 1 male : female ratio, although of course they have used wider definitions (Chapter 6 in Taylor, 1986). If this holds for hyperkinetic disorder, then the prevalence rate should be approximately halved to apply to all children. Furthermore, the age range studied here could be one of particularly high risk for the disorder: many older children will probably have developed and no longer display the disorder, and many younger children will not be in difficulties because school is not yet making demands upon their attention and self-control. It is true that some studies have suggested rather little in the way of age changes in levels of hyperactivity (Goyette *et al.*, 1978, Taylor & Sandberg, 1984), but these studies are based solely on questionnaire ratings that could well be less sensitive to age effects. Finally, the area studied was one marked by many of the inner-city problems that are associated with high rates of child psychiatric disorder (Quinton, 1988).

Nevertheless, and even after making allowance for age and gender effects, the prevalence is obviously greatly higher than the rate of about 12 per 10 000 who were in practice diagnosed as hyperkinetic in another part of inner London (Taylor, 1986, p. 11). Most such children are not referred to psychiatric or paediatric services. When they are referred, concern about their non-compliant behaviour and poor educational progress is likely to dominate descriptions by teachers and parents and could lead clinicians to underestimate the significance of the hyperkinesis and to diagnose conduct disorder instead.

Relationship to other studies

There are some apparent discrepancies between our findings and other published investigations, and the methodological differences need to be understood to avoid a sense of contradictions in the literature.

Widely different prevalence rates have been obtained by other investigators. This study identifies cases of hyperkinetic disorder on the basis of operational criteria using validated cut-offs on standardized measures. It is therefore not comparable with questionnaire-based surveys that have given higher values for pervasive hyperactivity. Schachar, Rutter and Smith (1981) reported from the Isle of Wight study that 2.2 per cent of all their 9–10-year-old children showed pervasive hyperactivity, and one may calculate a figure of 3.0 per cent for their boys. This contrasts with our present value, using the same questionnaire measures of 9.0 per cent; and emphasizes our previous point that the urban population studied here was likely to be a high-rate group. The urban-rural difference is particularly strong for chronic disorders of early onset (Quinton, 1988), and for disorders with hyperactive symptomatology (Boyle *et al.*, 1987).

Most epidemiological studies applying psychiatric diagnostic criteria of hyperkinetic syndrome have identified too few cases to give a prevalence rate; and in any case the low reliability of clinical psychiatric diagnosis forbids any generalization across studies. Vikan (1985), for example, suggested a rate of 0.4 per cent, but it was based on only three cases. Similarly, the Isle of Wight studies identified two cases in a screened population of more than 2000 (Rutter, Tizard, & Whitmore, 1970). The closest study to the present one in method is the two-stage investigation of children with 'minimal brain dysfunction' (MBD) in Sweden reported by Gillberg, Carlstrom, and Rasmussen (1983). They used operational criteria of ratings of behaviour by psychiatrist, neurologist, and parent. Their overall rate of 3 per cent for MBD with generalized hyperkinesis is, as stated by the authors, an overestimate, taking no account of persistence of symptoms over time. Their estimate for the rate of hyperkinetic disorder in those who had shown attention deficit in the screening stage of the study (the most comparable figure to our own) gives a population prevalence of 0.9 per cent for both sexes combined. Their sex ratio of 2.6 : 1 implies a point prevalence of about 1.3 per cent in males; which is compatible with the present figure of 1.7 per cent.

Epidemiological studies investigating ADDH do of course find high prevalence rates, as did our own application of ADDH criteria. Most such studies are in fact based on questionnaire ratings rather than diagnostic criteria, and the resulting rates obviously depend heavily on what cut-off happens to be taken in each study as well as on levels of deviance. However, Shekim and co-workers (1985) found a prevalence rate of 14 per cent on the basis of structured interviews and DSM–III criteria in a community sample of 9-year-old children, or about 19 per cent for boys only; as compared with our value of 17 per cent based only on the structured interview and DSM–III criteria.

In short, when like is compared with like there is substantial agreement about what is being measured.

Nor is there a serious contradiction between the neuropsychological dysfunction that we have found in the hyperkinetic disorder and the general lack of positive findings based upon previous comparisons of the DSM–III diagnoses of conduct disorder and ADDH (Werry *et al.*, 1987). The definition of hyperkinesis used here is both more restrictive, in that it defines fewer cases, and more specific, in that it is based upon inattention and restlessness only, rather than the impulsive and rule-breaking behaviours that help to define ADDH. Further, we were able to find definitely non-hyperactive cases of conduct disorder, which has proved difficult in clinic studies.

The associations of hyperkinesis in this general population sample are similar to those reported from case-record studies of handicapped people. Jenkins and Stable (1971) matched severely hyperkinetic patients with controls from a case register of intellectual subnormality; they found a higher rate of motor disorders and language delay. Thorley (1984) also found language delay and clumsiness to be more common in severely hyperkinetic children—many of whom were intellectually retarded—than in patient controls matched for IQ and age. It seems possible that the same disorder studied here also affects the handicapped, but with a higher rate. British clinical practice should consider extending the numbers of normally intelligent children who are recognized as having hyperkinetic disorder.

Do our 'cases' really have hyperkinetic disorder?

The prevalence rate suggested here is much higher than most UK psychiatrists suppose, even though it is considerably lower than that assumed in the US. British readers might therefore object that our methods have identified minor and non-specific levels of hyperactive behaviour, that we applied different standards during interviewing from those we used in the clinic, and that the resulting category is not valid as a disorder even if the clinical grouping is valid.

The quick reply to the objection is that our methods of identifying cases are the most rigorous yet used, and that the validity of the category is supported by its specific associations with independent variables. A complete reply to the objection would need to include a demonstration of a different course in the children with hyperkinesis, and these data are obviously not yet available. We have found that clinically diagnosed children with hyperkinetic disorder have a less favourable outcome in adolescence, even than matched psychiatric controls (Thorley, 1984); but the sceptic could still argue that our cases from the population may not be comparable with cases seen at psychiatric clinics. We must therefore consider the extent to which the cases identified in this study are dissimilar from the clinic cases.

The population rate of clinically diagnosed hyperkinesis has not been studied previously. However, the diagnosis of hyperkinetic disorder at British clinics is mostly made in children referred for problems of disruptive conduct (Taylor *et al.*, 1986*b*) and so one could roughly estimate the population prevalence from the prevalence of conduct disorder in the population and the proportion of those referred to a clinic with conduct problems who receive a diagnosis of hyperkinetic disorder. Isle of Wight data estimated a population prevalence of diagnosable conduct disorder at 4 per cent for all children aged 9–10 years, corresponding to approximately 6.4 per cent of boys (Rutter, Tizard & Whitmore, 1970). The figure should be increased for an inner-city area, but it is not clear by exactly how much (Rutter *et al.*, 1975): proportionate calculation on the basis of questionnaire scores suggests a figure of roughly 8 to 10 per 100. Our clinical researches (Taylor *et al.*, 1986*b*) have suggested that about 1 in 5 of boys referred to clinics with disruptive conduct should be diagnosed as showing hyperkinetic disorder (although in practice, of course, they are not). If the same proportion held in the population—a large assumption—then the population prevalence of hyperkinetic disorder should be around 1.6 to 2.0 per 100—which agrees with present findings. Such a calculation does not suggest that the research criteria were over-inclusive.

The argument above is fragile, and depends upon too many untested assumptions. It might seem better to report the diagnosis of a psychiatrist upon our current subjects. Indeed, one of us (E.T.) reviewed all data collected for the children with hyperkinetic disorder, those with non-hyperactive conduct disorder, and the controls, to make diagnostic judgements and to ensure that exclusion criteria were observed. This process, which was blind to diagnostic group membership, gave diagnoses of hyperkinetic syndrome in 14 out of 16 identified as such by the research criteria and in none of the other groups. However, this is not really a validation of the research criteria, for it was based, *inter alia*, upon the same information that led to the research scores. Furthermore, the diagnosis would not necessarily have been applied by other British psychiatrists. But the finding does suggest that a group close to our clinical concept of hyperkinesis has been identified. Direct comparison between this group, and previously studied cases from clinics, is possible for a few key measures.

The community cases of hyperkinesis do of course score more highly than controls on the same measures of hyperactivity which distinguished clinic cases. The major differential associations of the hyperkinetic subjects at clinics included lower IQ, more neurological signs, and more language delay. These were assessed in just the same way in the population and the clinic studies: the two types of case were comparably impaired. Thus, the hyperkinetic cases in the population had a mean IQ of 87 (compared with 85 in the clinic cases), a neurological score of 16 (21 in the clinic cases) and a reading comprehension 11 months behind that predicted (10 months in the clinic cases). Forty-four per cent had a history of language delay (53 per cent in the clinic cases). All this suggests that the same type of disturbance is being recognized.

The severity of disturbance was probably less in the population 'cases' by comparison with those seen at the clinic. The mean hyperactivity-inattention factor score of the Conners CRS was 1.9 (2.6 in clinic cases); the mean hyperactivity score of the PACS was 1.4 (1.8 in clinic cases). For conduct disorder the corresponding figures were 1.1 v. 1.4, and for emotional disorder 0.7 v. 0.8. There is a similar ratio between community and clinic for the rating scale of hyperactivity (0.73) and the interviewer's judgements (0.77); so it is very unlikely that the interviewers were adopting different standards of judgement in the epidemiological and the clinical studies.

The cases from the epidemiological study also tended to have better scores on the paired associate learning test than did the clinic cases (56 per cent errors v. 70 per cent), more correct responses in the continuous performance test (58 per cent v. 38 per cent), fewer impulsive errors in the same test (8 per cent v. 12 per cent), and a longer digit span (7 v. 4).

In short, we conclude that the assessment measures had the same significance in this population study as in clinic studies; that the community cases and clinic cases show the same type of disorder; and that the community cases, as a group, have a somewhat milder degree of disorder. There is also a good deal of overlap. While we cannot assume that the same adverse prognosis known for clinic cases will apply to these community subjects, we consider that they are at risk. Study of their outcome should be a priority for the future.

SUMMARY

Boys with hyperkinetic disorder were identified by a two-stage procedure in a community sample, involving parent and teacher screening-questionnaires followed by detailed interview, rating scale, and observational measures. The prevalence of disorder was approximately 1.7 per cent in 7- to 8-year-old boys (excluding those with known, severe learning disability). The disorder was separable from non-hyperactive conduct disorder by its early onset and frequent association with cognitive impairment, motor clumsiness, language delay and perinatal risk. Like those with conduct disorder, affected children were often living in families characterized by high levels of critical expressed emotion, low levels of coping, inconsistency between parents, and depressed mothers. The non-hyperactive but conduct-disordered children were specifically characterized by frequent marital discord between their parents and high rates of conduct disorder in first degree relatives.

8 Discussion

This monograph opened with questions about the nature of childhood hyperactivity. There were serious conceptual problems, not just technical difficulties in defining cases. The research that we have described has tested the idea of a hyperkinetic disorder and found it to be a useful way of describing psychopathology. Hyperkinetic children differ in important respects from controls, and not only from normal controls but also from children with non-hyperactive conduct disorder (Chapter 7). Hyperkinesis is a condition of early onset, associated with multiple cognitive and motor developmental delays. The presence of the disorder predicts features that did not define it. The differences between groups do not come only from the hyperkinetic children being more severely disturbed, but also from the type of disturbance they show (Chapter 4). We can therefore return to the original issues about the significance of hyperactive behaviour, with some confidence that they are truly important questions.

COMPONENTS OF HYPERKINETIC DISORDER

Over-activity

Locomotor over-activity is indeed a part of the disorder. This is not completely self-evident, even given that restlessness and fidgetiness are part of the definition. It would be possible for 'restlessness' and 'fidgetiness' to describe a change in the patterning of activity rather than an actual excess of movements made. However, direct observation and mechanical recording make it clear that the children who are rated as hyperactive are making more movements than others. Furthermore, all the situations investigated in the interviews give higher scores in the hyperactive children. Set tasks in the classroom and those allowing children a choice of activity, mealtimes at home, quiet play, watching television, and playing with other children, all show higher activity of children rated as hyperactive.

The North American literature on ADHD has also shown that ratings of hyperactivity can usually be validated against objective activity measures (Taylor, 1986). More than this, prolonged measurements of activity by Porrino and co-workers (1983) have shown that hyperactive children make more movements than others even during the night-time. This demonstration, that high activity in the hyperactive could not be reduced to inattentive off-task activity, has helped the shift from 'Attention Deficit Disorder' in DSM–III to 'Attention Deficit-Hyperactivity Disorder' in DSM–III-R.

Inattentive behaviour

It is clear that hyperactive children behave inattentively in many different situations, that their inattentiveness is present in the activities they choose at home as well as in tasks set by schoolteachers, and that it can be objectively recorded by an outside observer. 'Inattentiveness' of behaviour means a lack of persistence in activities ('short attention span'), and much off-task behaviour, changes of activity, and orienting to irrelevant aspects of the environment ('distractibility'). Inattentiveness therefore deserves its central place in descriptions of hyperactivity.

Some reservations remain. Inattentiveness is not a sufficient description: there are many ways of being inattentive in behaviour. This investigation has identified three types: the hyperactive pattern of pervasive restlessness and distractibility; the rather inert and apathetic picture seen in the 'pure inattentiveness' group considered later; and the distractibility of some anxious and miserable children. It should be useful to compare these types with more microscopic observation. They may even need different kinds of remedial action. Already it is clear that the types are too different to be muddled together into a catch-all category of attention deficit disorder.

The presence of inattentive behaviour, even when it is hyperactive in type, should not be the sole consideration in categorizing children. It is symptomatic of several types of problem. Some such children, as our questionnaire survey made clear, show an emotional disturbance. Hyperactivity can be a marker to severity of conduct disorder (Chapter 2). Hyperactive behaviour is often seen in autistic children and can be the dominating presentation in other types of pervasive developmental disorder. We have argued elsewhere, on circumstantial grounds, that hyperactivity in autistic children has a different meaning (see Chapter 8 in Taylor, 1986), but the question is by no means settled.

Inattentiveness is also a confusing description, because the word can be taken to mean either a type of behaviour or a score on a psychological test or an inferred psychological process. Although the behaviour of hyperactive children is plainly inattentive, experimental analyses of the supposed deficit have not yielded anything that to a theoretical psychologist would suggest an impairment of attention (see Chapter 4 in Taylor, 1986; Prior & Sanson, 1986). We have, of course, reported a pattern of impaired test scores in this book. However, though the tests were supposed to be measuring aspects of attention, the nature of the children's impairment did not suggest that central processes of attention were necessarily involved. Chapter 4 in the present monograph argued this point at more length: the clearest evidence came from the test of central against incidental learning in the children identified as hyperactive in the original questionnaire screen. Central learning was impaired in the most hyperactive group. We had hypothesized that incidental learning would be correspondingly better on the supposition that the children might show an ineffective deployment of their intellectual resources. In the event, however, incidental learning was no better in the hyperactive children and the ratio of central to incidental learning was not decreased.

Most other studies that distinguish central from incidental learning in children with ADHD have failed to find evidence for a deficit in selective attention (Douglas, 1988). An interesting study by Ceci and Tishman (1984) is the exception: it reported that hyperactive children had a lower ratio of central to incidental learning. This study differed in several respects from the rest of the literature: for example, it used recognition rather than recall to measure learning, and the stimulus materials were possibly more likely to attract children's interest. It is therefore possible that variation of technique could have led us to a similar finding. However, the technique we used led to a lowered score on central learning: this lowered score could not be accounted for by a failure of selective attention.

Failure to sustain attention is not the reason for poor test performance either. Cognitive impairment was present in the shortest test (the digit span) yet not in the longest (the continuous performance test). Errors were as likely at the beginning as at the end of tests showing impairment. In our clinical studies, the continuous performance test has shown worse scores in hyperkinetic children, just as it has in a number of North American studies of clinical ADDH (Werry *et al.*, 1987). But even when a deficit has been shown, it has not proved to be a function of test length, nor to increase as the test goes on. The ability to maintain test performance over time is not the crucial process limiting achievement.

The term 'attention deficit' is therefore something of a misnomer for the condition we are studying. This is no contradiction of the earlier statement that hyperactive children's behaviour is impersistent and distractible. It simply emphasizes the distinction between behaviour and cognition; and that the reason for the link between hyperactivity and poor test performance should be reconsidered—as it will be later.

Impulsiveness

Chapter 1 explained why impulsiveness was not part of our definition of hyperactivity. It means too many different things to different people. One popular part of the idea is moral: that hyperactive children break rules because of poor impulse control rather than because of any settled intention to flout them. We could not test this idea: in fact it is hard to translate into testable terms. If the impulse is the explanation for behaviour yet is only known to be present because of the behaviour, then the argument is circular and the concept best avoided.

Nevertheless, several of our findings point to other aspects of impulsiveness. Careless, disruptive behaviour in the classroom is prominent in hyperactive children, but not characteristic of them (Chapter 4). At home, the high frequency of minor accidents in the hyperkinetic group supports the association we found previously in a clinically defined group (Chapter 6).

The 'Matching Familiar Figures' test elicits variation in the speed of responses. Swift, inaccurate responses can be interpreted as one form of 'impulsiveness'. We found that boys with hyperkinetic disorder are quicker in their responses, as well as less accurate. There is a strong correlation between the speed of the response and the number of errors made. Indeed, the correlation is high in controls (0.68) as well as in pervasively hyperactive children (0.65). It is therefore possible that an over-rapid tempo of responding contributes to a poor score on this test and to a wider difficulty of learning.

Full interpretation of the meaning of the rapid inaccurate responses is a little complicated. It may be intuitively obvious that 'of course' a hyperactive child will react more quickly, but the evidence is more conflicting than the common-sense view admits. In studies both by Firestone and Martin (1979) and Sandberg and co-workers (1978) it was found that pervasively hyperactive children made more errors on the Matching Familiar Figures Test than did controls with other types of disordered behaviour—but the time taken by the child was not less than that of the controls in either study.

The factors determining how long a child takes to respond are quite complicated. Traditional psychological tests of reaction time have found that children with ADDH have longer latencies when the pacing of the test is controlled by the experimenter (Sykes *et al.*, 1973) and during automated tests (Klorman *et al.*, 1988). It is clear that the children are not simply speeded up; to the contrary, they may even take longer to process simple information. Sergeant (1981 & 1988*a*) has examined the idea of impulsivity considered as inefficient information processing with fast, inaccurate control. He finds that children with ADDH can and do make fast, accurate responses on occasion. Indeed, when they respond slowly, they are more likely to be wrong than when they respond quickly. He concludes, correctly, that they are not impulsive in their cognitive processing.

This may seem like a flat contradiction of our present results, but it is not. True, we have been able to find a clear relationship between responding quickly and responding incorrectly. Within individual subjects, correct responses were made more slowly than incorrect ones. But the psychological meaning of the long-latency response in the MFF test (average 9.6 seconds in our pervasively hyperactive groups) is different from that of the short-latency response in Sergeant's information-processing test (average 1.5 seconds in his ADDH group). The long latency of the MFF is largely determined by the children's decision about how long to spend, not by their capacity to go quickly. A long latency is not always an index of the amount of effort spent—the time may be spent in fidgeting or amusing oneself as well as in figuring out the right answer—so teaching children to go slowly would not be a sufficient therapy. But, if one wants to get the answer right first time, then the decision to take a long time over it is adaptive. In this sense, the children are regulating themselves inefficiently when they respond quickly and make mistakes.

On the other hand, the rather pejorative word 'inefficiency' comes from an adult-centred view about what the child should be doing. It does not follow that the children shared our value scheme, or specially wanted to get the answer right first time. Their decision-making may well have been based on different premises. This is not to say that they were being disobedient or indifferent to the examiner's instructions and response. On the contrary, the children who were inaccurate in their replies did not tend to show defiant or oppositional behaviour (Chapter 4). But, the children were able to go on guessing after an error. They could have decided upon a strategy of quick, multiple guessing and so hit upon the correct answer as rapidly as by laborious thought. Perhaps hyperactivity in such situations tends to be maintained because it is effective in increasing the rate of gain of reward. Even if this is not the case, 'impulsive' behaviour still cuts down the time spent in delay before action; and delay could be unpleasant for children who are hyperactive. Rapport and co-workers (1986) have found that children with ADDH prefer immediate gratification. Future research could profitably make a systematic investigation of the effects of incentives and delays in rewards in the determination of response time and accuracy.

As with attention, so with impulsiveness: we have needed to draw a clear distinction between behavioural observations and the cognitive processes that could be linked to them. Impulsive behaviour—badly timed, intrusive on others, accident-prone—certainly exists in children with behaviour problems, but it is just as common in non-hyperactive conduct disorder as it is in pervasive hyperactivity. This sort of impulsive behaviour does not have discriminative validity, so should not be used in defining hyperkinetic disorder or ADHD (although it may well need to be recognized and treated). On the other hand, hyperactive behaviour is specifically linked to an unregulated and often maladaptive style of behaviour during problem-solving; a style that reflects children's decision-making rather than a deficiency in processing information. This is the important idea behind the statement that hyperactive children are impulsive.

Temperament or disorder?

The next issue about the nature of hyperactivity is whether it should be conceived as the extreme of a temperamental dimension or as a psychiatric syndrome; or rather (since both may be true) which idea works best in practice.

Different aspects of our results support both ideas. The distinction between dimension and category is a false antithesis (see Chapter 1). The distribution of scores on questionnaire ratings of hyperactivity is continuous: there are progressively fewer cases at successively higher levels of severity (see Chapter 1 in Taylor, 1986). This clearly supports the view that there is a continuum of hyperactive behaviour shading into normality.

On the other hand, the associations between hyperactive behaviour and neurodevelopmental impairment do not seem to be comparable at all points of the range. Only the more severe cases, of hyperkinetic disorder defined by research criteria, showed lowering of IQ, motor clumsiness, and perinatal adversity. The discontinuity at a higher level of the range implies that the extreme of the dimension may be qualitatively distinct in its determinants. This is not a contradiction, but a frequent type of finding in neuropsychiatry. Low IQ, for instance, is at the same time a low score on a dimension that predicts a continuum of later social impairment, and a score that identifies (at levels below 50) a group with distinct biological causes that play little part in determining individual variation within the normal range.

Hyperactivity has not been satisfactorily reduced to other dimensions of temperament. Some kinds of personality theory seek to map disorders on to a few fundamental dimensions. Eysenckian theory, for example, might describe the restless and changeable behaviour of hyperactivity as an extreme of extraversion. This has not been upheld in children: we have found no correlation between hyperactivity and 'extraversion', but significant though small correlations with 'psychoticism' and 'neuroticism' (Chapter 5 in Taylor, 1986). Cloninger's (1986) theory could be used to describe hyperactive children as low in 'harm-avoidance' and high in 'novelty-seeking' and 'reward-dependence'; but it would have to be shown that this combination was specific for hyperactivity, and at least as predictive, before replacing the descriptions.

Genetic evidence might well be helpful in clarifying whether the familial links are with a trait of activity or attention or with a psychiatric disorder. There is no real doubt that something connected with hyperactivity is inherited. Comparisons of concordances between monozygotic and dizygotic twin pairs make it clear that there is a genetic contribution to the dimension of hyperactivity (Buss & Plomin, 1975), as to other temperamental traits. It is much less clear whether the hyperkinetic disorder shows increased genetic loading for that and other psychiatric disorders in families. A recent twin study has used the same definition of pervasive hyperactivity employed in our first stage and demonstrated a genetic contribution (Goodman & Stevenson, 1989). It suggested that pervasive hyperactivity was linked to situational hyperactivity in the classroom, but that hyperactivity confined to the home might be somewhat different. This analysis, however, could not take account of different reliabilities by school and home raters; and did not contain enough deviant subjects to examine the inheritance of the hyperkinetic disorder as identified by this study. Genetic strategies should be used more frequently in studying the basis of hyperkinesis.

A temperamental account of hyperactivity carries other implications. The original proponents of the idea of temperament laid emphasis on its being an aspect of individual variation that was neither the result of motivated, learned behaviour nor the result of disability. The association between hyperactivity and test performance emphasizes that we are dealing not only with a difference but with an impairment.

Theories about temperament stress the interplay of constitutionally determined variability with other risk and protective factors, rather than the predetermined unfolding that is implied by simple disease models. This view, while it may do scant justice to the complexity of some medical models, is in keeping with our own findings. Both the short-term follow-up of the pervasively hyperactive groups in this study, and the long-term controlled follow-up of severely affected clinical cases by Thorley (1984) indicate that the outcome varies with other aspects of the children (aggression, IQ) and their families (high expressed criticism, maternal depression, low social class) at least as much as with the severity of the hyperactivity.

In summary, we need to think both of a dimension of individual variation ('hyperactivity') and also of a syndrome of disorder of which hyperactivity is part ('hyperkinetic disorder'). The homogeneity of the disorder, its overlap with other conditions, aetiology, and pathophysiology, are worth a good deal more study.

Issues in the definition of a hyperkinetic disorder

We have argued that a disorder ('hyperkinesis') needs to be recognized as well as a dimension ('hyperactivity'). Inattentive restlessness gives the definition of hyperactivity; but the recognition of a valid disorder needs more than simply the presence of some degree of hyperactivity.

The DSM–III definition of ADDH is a helpful starting point. It has many strengths that do not need to be rehearsed here: in particular it is so explicit that it can be applied systematically in this kind of survey. Its weaknesses, however, are too great for it to be maintained as a disorder. As discussed in Chapter 7, it applies to about one boy in every six. Of course, one must not dismiss the idea of a psychiatric condition simply because it is very common. In this instance, however, not only is ADDH very common but most boys who meet the criterion are not socially impaired and are not even encountering problems in educational progress. This is not because we have unusual ways of recognizing ADDH, for our diagnostic judgements agree with a North American research team (Prendergast *et al.*, 1988). Nor have we diluted the ADDH category to the point of losing its validity. The boys with ADDH are indeed different from controls in their activity and attention. Their attention test scores could be rather specifically impaired, for their worse performance on tests such as the Matching Familiar Figures is not accompanied by any shift in IQ and is still significant even after IQ is covaried. In these respects our findings are similar to the case-control comparisons reported from clinically referred series in other countries.

The relative normality of most boys with ADDH (unless they meet criteria for hyperkinetic disorder as well) should be a caution against the diagnosis. In clinical practice, ADDH could well seem to be a serious risk factor for later development. The children who receive the diagnosis have been referred for help, so they are much more likely to show social and educational impairment. Many of them may well meet criteria for hyperkinetic disorder; even those that do not are likely to have something else wrong with them. Their poor social outcome has often been documented (e.g. Weiss & Hechtman, 1986); sometimes a very poor outcome has been reported (e.g. Satterfield *et al.*, 1982). Nevertheless, the implication of studying ADDH in a representative population is that the presence of ADDH is not likely to be the main risk factor; the other factors leading to referral are of more import. Research into the basis of ADDH should therefore stop relying upon comparisons between clinic cases of ADDH and controls from the non-referred population. Conclusions drawn solely from such designs are likely to be seriously misleading.

The weakness of the ADDH definition is not only that it includes relatively normal behaviours, but also that it lacks discriminative validity (Rutter, 1989*b*). Werry and co-workers (1987) reviewed the studies that could yield comparisons between different DSM–III diagnostic groups, and had to conclude that differences could be important but had not been demonstrated. Their own empirical study provided the best comparison between DSM–III diagnoses in a clinical population, and was based on a structured interview with parents to make the diagnoses (Reeves *et al.*, 1987). They found few differences between diagnostic groups, and those that they did find were nearly all between anxiety disorders and the externalizing behaviour disorders: conduct and attention deficit disorders were not differentiated.

One probable reason for the lack of valid differentiation between the disruptive behaviour disorders of the DSM–III classification emerges from the present study. While defiance and inattentive-restlessness can be readily distinguished, the behaviours of impulsiveness are common to both. Impulsive behaviour is a prominent part of the definition of ADDH. The result is likely to be a loss of discriminative validity because of the emphasis on a class of behaviours that may be important for prognosis but are non-specific concomitants of disorder.

The simultaneous presence of several types of pathology is very frequent in clinically referred cases. In this population study the overlap is less marked: emotional disorder and conduct disorder are independent constructs. Even in the population, however, hyperactivity is quite strongly correlated with conduct disorder; and, to a lesser extent, hyperactive children tend to show anxiety symptoms at school (Chapter 2). This co-morbidity needs to be understood more deeply before separate disorders are defined. It is quite possible to make multiple diagnoses and recognize all the dimensions of disturbance that are present; but the procedure is complex and will fail to recognize any special properties of mixed disorders.

In practice, multiple and simultaneously present disorders are taken as raw material for a higher-order classification. One can then focus on ADDH without conduct disorder or emotional disorder; ADDH with emotional disorder but not conduct disorder; and so on through the possible permutations. However, there is a risk of adopting an arbitrary classification, and the scheme becomes very awkward when the number of lower-order categories (or dimensions of disturbance) rises above three. Several patterns of disorder can obviously be recognized from their profiles of behavioural disturbance. The question is, which are useful?

When the different stages of this study are considered together, several patterns should be recognized. The detailed measures tested and confirmed the pattern of hyperkinetic disorder, characterized by pervasiveness, severity, and persistence of hyperactive behaviour. A second pattern is a non-hyperactive conduct disorder, of later onset in childhood.

Some other disorders are suggested, but will need further exploration and testing. A mixed category of conduct and emotional disorder has been part of the World Health Organization's classification scheme for some years, and is supported from the different pattern of emotional symptomatology found in the mixed disorder from that seen in uncomplicated emotional disorder (Chapter 3). A group of emotionally disturbed children emerged in a cluster analysis of disruptive boys referred to psychiatric clinics (Taylor *et al.*, 1986*b*). Emotional disturbance at school predicted a poor response to methylphenidate in a group of disruptive boys of whom many were hyperactive, which suggested that the presence of anxiety might be a meaningful distinction within the group (Taylor *et al.*, 1987). 'Affective conduct disorder' therefore seems to have earned its place in classification. It ought to be useful to study disturbances of attention in emotionally disordered children and to compare them with the disturbances seen in the hyperkinetic disorder.

Hyperkinesis also needs to be distinguished from inattentiveness without hyperactivity (AD) (see Chapter 4). The two types of behavioural disturbance are so different that it seems very dubious that they can be considered as variants of one condition. The AD group has a different pattern of symptoms and predicts differently. It, alone among the groups from the questionnaire screen, is associated with a lower IQ and delayed language development; and, unlike the hyperkinetic group, it is not associated with inattentiveness test scores once IQ has been allowed for. Its association with low social class is similar to that of children with mild intellectual retardation and unlike that of children with hyperkinesis. Possibly the early ratings of 'poor housecraft' could be evidence for an impoverished early environment. It is also dissimilar from hyperkinesis in its lack of association with disturbance of relationships between family members. It is not likely that inattentiveness represents an early stage of the development of hyperkinetic disorder: the early antecedents of hyperkinesis are somewhat different, and AD did not lead on to hyperactive behaviour by the time of the second stage. AD seems to arise and develop separately from hyperkinesis. There does not seem to be much need to identify it as a specific psychiatric disorder, since it is essentially a group with general cognitive impairment and not much else.

Other patterns are still speculative, and outside our present scope. For instance, children with a schizophrenic inheritance may have a characteristic impairment of attention. But it is evident that many disruptive children do not neatly belong in any of the groups set out. Those with lesser degrees of hyperactivity—more than the non-hyperactive conduct disorders but less than the hyperkinetic—seem likely to fit better with the non-hyperactive conduct disorder on this evidence. The area left rather obscure is that of children with severe but situation-specific hyperactivity. The suggestion from a recent twin study is that hyperactivity confined to the home may be separable from other forms (Goodman & Stevenson, 1989). This is likely to be a fertile field of enquiry for the future.

These considerations persuade us that a discriminating 'diagnostic' approach is valuable for clinicians and researchers. The problems that we have identified in the DSM–III scheme have not been entirely resolved by the revision into DSM–III-R. Certainly DSM–III-R makes it clear that hyperactivity (rather than attention deficit alone) is the core of the disorder, and this is a welcome move. But the excessive numbers of children identified—because of the inclusion of some unremarkable types of behaviour as symptoms—and the lack of discriminative validity stemming from the emphasis on impulsivity have not been altered. Further, the issue of conflicting information has if anything been muddled, for the clear rule of DSM–III (to give preference to teachers' observations) has been replaced by a lack of guidance. The idea of a hyperkinetic disorder that emerges from the present work is closer in spirit to the ICD-10 definition (WHO, 1988), Barkley's (1982) criteria, and the criteria put forward by Sergeant (1988*b*). The research definition in ICD-10 is particularly close to that used here, though the exact cut-offs are not validated; it is given below:

The research diagnosis of hyperkinetic disorder requires the definite presence of abnormal levels of inattention and restlessness that are pervasive across situations and persistent over time, that can be demonstrated by direct observation, and that are not due only to other disorders such as autism or affective disorders.

Eventually, assessment instruments should develop to the point where it is possible to take a quantitative cut-off score on reliable, valid and standardized measures of hyperactive behaviour in the home and classroom, corresponding to the 95th percentile on both measures. Such criteria would then replace 1 and 2 below.

All the following must be present:

1. Demonstrable abnormality of attention and activity at home, for the age and developmental level of the child, as evidenced by at least three of the following attention problems:

   (a) short duration of spontaneous activities
   (b) often leaving play activities unfinished
   (c) over-frequent changes between activities
   (d) undue lack of persistence at tasks set by adults
   (e) unduly high distractibility during study e.g. homework or reading assignment

and by at least two of the following activity problems:

   (a) continuous motor restlessness (running, jumping, etc.)
   (b) markedly excessive fidgeting and wriggling during spontaneous activities
   (c) markedly excessive activity in situations expecting relative stillness (e.g. mealtimes, travel, visiting, church)
   (d) difficulty in remaining seated when required

2. Demonstrable abnormality of attention and activity at school or nursery (if applicable), for the age and developmental level of the child, as evidenced by at least two of the following attention problems:

   (a) undue lack of persistence at tasks
   (b) unduly high distractibility, i.e. often orienting towards extrinsic stimuli
   (c) over-frequent changes between activities when choice is allowed
   (d) excessively short duration of play activities

and by at least two of the following activity problems:

   (a) continuous and excessive motor restlessness (running, jumping, etc.) in situations allowing free activity
   (b) markedly excessive fidgeting and wriggling in structured situations
   (c) excessive levels of off-task activity during tasks
   (d) unduly often out of seat when required to be sitting

3. Directly observed abnormality of attention or activity. This must be excessive for the child's age and developmental level. The evidence may be any of the following:

   (a) direct observation of the criteria in 1 or 2 above, i.e. not solely the report of parent and/or teacher

   (b) observation of abnormal levels of motor activity, or off-task behaviour, or lack of persistence in activities, in a setting outside home or school (e.g. clinic or laboratory)

   (c) significant impairment of performance on psychometric tests of attention

4. Does not meet criteria for Pervasive Developmental Disorder, Mania, or depressive or anxiety disorder

5. Onset before the age of 6 years

6. Duration of at least 6 months

7. IQ above 50

ASSOCIATIONS AND MECHANISMS OF HYPERKINESIS

The numbers with hyperkinetic disorder yielded by our study design are low, because of the infrequency of the condition; but they are not likely to be unrepresentative and spurious associations are correspondingly unlikely. The consequent lack of power in statistical tests means that we may have overlooked associations that are in fact present. Because of the small numbers and lack of power we applied a relatively undemanding level of statistical significance (0.05). In spite of this and the many comparisons made, we do not consider that Type 1 errors are likely; since the main findings, of developmental and cognitive associations with hyperkinesis, are based upon several corroborating, independent sources of evidence; and composite scales give the same results as individual measures. Nevertheless, isolated positive findings (such as those specific to conduct disorder) should be seen as needing replication.

   The link with perinatal risk could have several causes. The most obvious—that perinatal adversity injured the developing brain—is not necessarily correct. The 'risk' factors were of a type usually considered, in present knowledge, to be minor, and the reason for special care was usually jitteriness or delay in establishing respiration, or evidence of neonatal hypoxia in a baby of normal birth-weight and unremarkable delivery. These seem more likely to be the result of a constitutional abnormality than the cause in themselves of brain damage, but the contemporary information is not full enough to allow for much confidence either way. Neonatal hypoglycaemia, for instance, cannot be ruled out. A large study focusing on the later outcome of children with birth complications has suggested that 'hyperkinetic-impulsive' behaviour at age seven is predicted only weakly by any complications of pregnancy and delivery; and that maternal smoking in pregnancy, low foetal heart rate during labour, and a small head circumference at birth are the strongest of the early factors (Nichols & Chen, 1981). This would support our view that factors in prenatal development, rather than injury during or after birth, are of most importance in the initiation of severe degrees of hyperactive behaviour. It is also likely that the experience of the baby's being at risk has an effect upon the system of family relationships (Corter & Minde, 1987). The association with neonatal problems, stronger in the present study than in others, comes from the use of a more restricted definition: it is not the result of retrospective falsification of recall or of an independent association with social disadvantage.

Cognitive impairment

The link between hyperkinetic disorder, behaviourally defined, and cognitive impairment is strong. It is specific in the sense that it does not hold for conduct disorder without hyperactivity—at least in children of this age. The cognitive impairment in hyperkinesis is not specific in the sense of being a deficit in a single test or inferred process. Indeed, as we have discussed above, it is in some ways misleading to call it an attention deficit at all.

This scepticism about any impairment of processes involved in information processing is not restricted to the specialized meaning that we have given to hyperkinesis. Review of the extensive, and often distinguished, psychological literature upon ADDH leads to the same conclusion. Deficits have, to be sure, appeared upon tests that require the processing of information. Douglas (1988) has been one of the many scholars to show that any deficits are very strongly dependent upon the experimental details of the situation, and to interpret this as the result of very high-level control processes of self-regulation and inhibition, rather than any failure in the elementary steps of perception attention and motor response.

Our finding of a worse performance on many tests, even those of the IQ battery, is not in line with the ADDH studies, and indeed we did not find it when the ADDH definition was applied. But we have already noted that many children, studied after referral to specialist centres because of ADDH, would probably meet the hyperkinetic disorder criteria as well. Referral is likely to be precipitated by other problems as well as the defining features of ADDH. This very process of referral can also create a false impression of specificity in the children's problems. Since specialized research clinics exist alongside other helping agencies, children with more generalized learning problems may well be referred preferentially to the latter. The view that ADDH represents a specific problem in learning could therefore be a self-perpetuating fallacy. Investigators using screened population samples rather than referred samples have found general rather than specific disabilities (McGee *et al.*, 1984, Sergeant, 1988*a*, Szatmari *et al.*, 1989).

What mechanisms might underlie this association between severe hyperactivity and psychological test score? We have already rejected the idea that the behaviour and the test score are different measures of the same fundamental process of attention. Other lines of evidence speak against the behaviour and the test performance being linked by direct and immediate causality. For one thing, the pattern of 'pure' inattentiveness, without hyperactivity, is also related to delays in intellectual development, yet the types of behaviour shown are really very different. For another thing, the response to medication seems to show that behaviour and cognition can quite readily be separated (Taylor, 1986, p. 207). Some children show a large behavioural change after stimulants, without concomitant changes in test scores. Some children show a lessening of hyperactivity after drugs (such as phenothiazines) that worsen test scores. This is not a final argument, for drugs typically have multiple effects, but it does argue against hyperactive behaviour being the direct and immediate cause of poor test performance, or vice versa.

One might think of a low IQ, language delay, clumsiness, and hyperkinetic disorder as all being aspects of a general immaturity: a child who is slow to develop will solve puzzles less well and might be less developed in attention and activity control. It is difficult to assess the notion fully because it is doubtful whether hyperkinesis simply represents an earlier stage of development. Two lines of evidence at least are against it: the very early onset of hyperkinetic disorder implies that affected children are recognizably different at an early stage, and are not simply prolonging early behaviour to an inappropriate age (Chapter 6). The swift response to stimulant (and other) medications is not a function of developmental level but of attention and activity (Taylor *et al.*, 1987).

Some other possible arguments against the immaturity explanation are specious. True, ratings of hyperactivity are usually not correlated with chronological age; but raters probably allow for the child's age in making their ratings, and objective measures of activity do tend to be age-related (Routh, 1980). True, hyperactivity can still be a handicap to development in adult life; but this does not exclude a developmental delay, especially as a delay in early life may bring many other secondary consequences in later life. Immaturity is not a sufficient explanation of the link between disturbed behaviour and low IQ, but could still play a contributory part. Testing the idea would be helped by biological measures of normal changes with age (such as the neurometric EEG equations, Prichep *et al.*, 1983), and by comparisons on psychological tests between hyperactive children and normal but younger children matched for test performance. It would then be possible to distinguish deviance from delay with more rigour than has been possible in a cross-sectional survey.

If a general immaturity of the brain does not explain the whole of the association between hyperkinesis and cognitive problems, then one must look to other causes in common or to a more complex developmental formulation. It looks as if pure inattentiveness has different antecedents from those of hyperkinetic disorder (see above), yet rather similar cognitive associations. This leads us to suggest that inattentive behaviour, whatever the cause, leads indirectly to a delay in cognitive development. There is some evidence for this: studies of normal children in infancy have suggested that the extent to which they prefer to look at novel stimuli, and sustain their interest in them, is quite a good predictor of IQ in later childhood (Berg & Sternberg, 1985). Indeed, it predicts more effectively than general measures of central nervous system development, such as the developmental quotient. This preference is in the line of development that goes from the orienting of attentive behaviour towards physically salient stimuli, to a preference for novel stimuli, to a preferential orienting to interesting and relevant aspects of the world. One of us has previously argued that hyperactivity is a disruption of this development towards complexity and organization of attentive behaviour (Chapter 4 in Taylor, 1986).

According to this idea, restlessness and impersistence are the early behavioural manifestations. They have the effect of making young children live in a less interesting world. If children do not sustain interest in new things, then they learn less about them and have less experience in understanding changes and less well-developed cognitive structures as a result. Hyperactive behaviour is aversive to adults, so may reduce the interactions that assist language, and other cognitive, development. In later childhood, the preference for novelty will peak in hyperactive children at an age when their peers have developed beyond it to a more organized and systematic style of exploration. By the time of entry into school (and of being studied by this project) they could be expected to be more obviously inquisitive and curious, but less effective in their curiosity, than their contemporaries. Direct observations of hyperkinetic children during interviews have indicated that this description applies quite well (Luk *et al.*, 1987).

It ought to be useful to apply experimental analyses as much to the behaviours involved in attending as to the stages of processing information. This will need closer description of orientation, search, scanning, and preferences, together with understanding their determinants through development and their relation to learning disabilities. It may also be necessary to understand how children form theories about what is relevant and interesting to guide their strategies of exploration.

The pattern of test performance is strikingly similar to that of the group of children who are identified behaviourally as showing an attention deficit yet are not hyperactive (Chapter 4). In this behavioural sense, it is reasonable to conceive of hyperkinesis as including an attention deficit: inattentiveness rather than hyperactivity gives the association with poor test performance. However, the behaviour of the hyperkinetic and attention-deficit groups is of course very different; as has also been noted in descriptive studies using different methodology, such as that of Lahey and co-workers (1984). It would therefore be difficult to regard hyperkinesis as the simple expression in behaviour of a deficiency in processes of attending. Perhaps intellectual impairment is a risk factor for hyperkinesis but requires the mediation of adverse family relationships; against this is the very early onset of the behavioural disturbance in hyperkinesis. The most plausible reason for the link between poor test performance and these two different types of inattentive behaviour is that the attending habits of children effectively alter the stimulation that they receive. A lack of persistent orientation to novel, complex, and interesting stimuli could result in an impoverishment of experience and therefore in a slowing of intellectual development. If intellectual performance were to be enhanced by treatment, then any improvement in attending behaviour would have to be maintained over long periods, and accompanied by an appropriately stimulating environment.

Family and social association

There are probably many initiating causes of inattentive and hyperactive behaviour. Genetic factors have already been noted on the basis of twin studies. The lack of a family history in our hyperkinetic cases should not be seen as excluding a genetic contribution. The ascertainment of affected relatives was based solely upon parents' report.

Though this is often a useful methodology, one must note that even for the probands parents often did not recognize a serious problem; when they did, the recognized problem was often not hyperactivity; and accordingly they would also tend to under-report the disorder in siblings or their own childhoods.

For those with non-hyperactive conduct disorder, the clustering of conduct-disordered relatives suggests that the familial transmission of conduct disorder is not mediated by genetic inheritance of hyperactivity. The transmission of non-hyperactive conduct disorder could be psychosocial or genetic on this evidence: one adoptive study has found a genetic component and indicates further research (Bohman *et al.*, 1982).

There are many psychosocial associations of hyperkinetic disorder, as indeed was suggested by the work of Gillberg and co-workers (1983) and Schachar and co-workers (1981). We did not confirm their suggestion that hyperactivity is linked to lower socio-economic status. The group identified by Gillberg was initially identified quite largely by inattentiveness of behaviour, and so this was probably the component linked to lower class. In our population the externals of social life played little role in the genesis of hyperkinetic disorder. Divorce, unemployment, separation, poor housing, and lack of education seemed by themselves, to take a small part. A poor marital relationship did not lead to hyperkinesis, though it may well have contributed to non-hyperactive conduct disorder in this study and to affective conduct disorder in our previous clinic survey (Taylor *et al.*, 1986a).

The signs of unhappy family life that we found are likely to be important for the later outcome of the children. They represented angry, inconsistent parental reactions, and the high levels of maternal depression could well have contributed to this hostile atmosphere. This pattern of family life was not specific to hyperkinetic disorder. Hyperactivity and defiance were equally strongly associated with poor coping and expressed criticism. The significance of the family relationships is therefore not so much a specific cause of hyperactivity as a factor strongly associated with persistence of disorder. The mechanism of the association could simply have been that persisting disorder was the cause of hostile family relationships. We cannot exclude this, but it is unlikely to be the whole story. The immigrant families, especially from the Indian subcontinent, tended not to be characterized by hostility, inconsistency of coping, or maternal depression. This is very likely to be a cultural phenomenon, and therefore not essentially the result of children's behaviour. Nevertheless, the children of immigrants from the Indian subcontinent tended more often to grow out of their hyperactivity during the period of the study.

It therefore seems likely that a hostile and unhappy family atmosphere helped to determine whether the children's disorder persisted. The possibility is sustained by the extent to which mental health in other family members and family socioeconomic status predict a better adult outcome for children with ADDH (Weiss, 1983) and hyperkinetic syndrome (Thorley, 1984).

The interactions between child constitution and family environment may operate in different ways at different developmental stages. We have argued that a deprived and unstimulating atmosphere makes it harder for a young child to develop good attending skills, and is therefore a risk factor for inattentiveness. It need not lead to hyperactivity; and indeed the earliness of onset of hyperactive behaviour, its links with neurodevelopmental delays, its known heritability and susceptibility to drug effects, all suggest that constitutional factors are the main initiators. However, the natural course of development will be for hyperactivity to decrease. This will be complicated by the reactions of other people to hyperactivity. It is likely to evoke hostile reactions, an unhelpful style of coping by adults, and unpopularity among other children. The extent to which this happens is modulated by other factors, including the mental health of other family members and culturally determined attitudes. The resulting negative relationships constitute a main part of the risk that affected children will develop persisting problems in social adjustment, remain generally hyperactive, and reach a point where they show hyperkinetic disorder and are all the more at risk.

RECOGNITION AND TREATMENT

The discussion so far has been in terms of the fundamental psychiatric understanding of the complex of problems called hyperactivity, and of future research. Our findings also carry strong implications for the clinical practice of child psychiatry. They suggest provisionally that the condition of hyperkinetic disorder needs to be recognized, and that it should be a target for treatment, more widely than at present in England.

There seem to be several reasons for the under-recognition of the condition. Parents are critical of the behaviours that their hyperkinetic children show, but do not identify them as a medical problem; or, if they do, they see it as antisocial (Chapter 7). This could be partly a result of advice received: the few parents who consulted their family doctor had received an impression that such problems were matters of child-rearing rather than of medicine. It could also result from wider cultural attitudes. During the period of this study the lay idea of hyperactivity, as communicated through the press and television, was of a disease caused by additives in food. The notion was rejected by clinical professionals, probably because of the lack of scientific support for it. The idea had also failed to catch on among most parents in the area where we were working, perhaps because it was seen as a rather precious and upper-class fancy. The existence of 'hyperactivity' probably tended to be rejected by those who could not accept the dietary theories of cause. It still needs to be stressed that the recognition and treatment of hyperkinetic disorder are entirely different issues from the frequency and significance of adverse reactions to food.

Teachers, by contrast, did indeed see problems for children with hyperkinetic disorders and were often seeking further advice from educational psychology. However, 'hyperactivity' and 'attention deficit' are not categories of special educational need and there are no special facilities for them; so the labels would largely be irrelevant and were not applied.

In the light of these circumstances and fashions, hyperactivity alone is unlikely to be a reason for referral. When children are sent to a specialized service it is likely to be because their problems are severe and include other disabilities. This is compatible with the high rate of conduct disorder in hyperactive children at clinics (some three in every four at the South London clinics we have described). The professionals seeing these cases diagnose the conduct disorder rather than the co-existing hyperkinesis (Prendergast *et al.*, 1988). This is not always wrong: Chapter 5 suggested that in most hyperactive children the determinants of short-term outcome are aggression, family discord, and maternal depression. But the practice of diagnosing only the conduct disorder neglects the presence of a subgroup of children with hyperkinesis, and the extent to which pervasive hyperactivity predicts a poor outcome in conduct-disordered children (Schachar *et al.*, 1981). We now need more detailed prognostic studies, but in the meantime severe hyperactivity should be noted and treated.

This does *not* mean that stimulant medication should start to be used as widely in the UK as it is in the USA. There is a range of treatments available; and, for most cases in a community, behavioural therapies and advice to parents will be the preferred means of health intervention (Yule, 1986). Stimulant drugs are valuable for the most severe cases that do not respond to simpler measures and therefore need specialist referral (Taylor *et al.*, 1987). The place of drugs in a comprehensive treatment scheme has been outlined by Schachar and Taylor (1986). Stimulants should probably be used more widely in specialist child-psychiatric and behavioural-paediatric centres; but there is no case yet for expanding their use into primary care. Indeed, the necessary monitoring of treatment requires specialist expertise. The successful treatment of hyperactive children needs more than the reduction of hyperactivity: interpersonal relationships are important targets for therapy.

Dietary treatments, by contrast, do not get much support from this survey, in that symptoms of physical ill health are not prominent in pervasively hyperactive children. We did not set out to test the dietary hypotheses rigorously, and would have needed experimental methodology and controlled trials to do so. Nevertheless, the infrequency of physical symptoms is in some contrast to the very high rates of headache and other symptoms in the chief study to assess dietary therapies (Egger *et al.*, 1985). Children responding to diet may therefore be a highly selected and small minority.

The need for services ought to be examined further by comparisons of treatment approaches over periods of one or more years. There can be no question about the short-term effect of the treatments mentioned above; but their efficacy in promoting healthy psychological development remains to be tested. It should be possible to provide the therapeutic services without any major expansion of resources. Children with hyperkinetic disorders are likely to consume a substantial amount of resources anyway. At this age the main input seems to be from teachers in mainstream classrooms, perhaps at the expense of other children. But, already, special educational placements are being envisaged for the future. Referrals to the health services are likely to increase as the children get older. We are therefore suggesting some change of focus in what the specialist services do, rather than a massive increase in case load.

Preventive services are also conceivable, but would probably require some extra resources and some further research to guide them. Early recognition of cases should not be too difficult. Health visitors and community physicians are already detecting behavioural and developmental impairment at an early stage (Chapter 6). It should be possible to improve on the process by clearer definition of the level of problems that should be grounds for early secondary prevention. At present, however, the problems often seem to be noted without leading to helpful action. Some of this could no doubt be remedied by better communication and co-operation between professionals. There could, for instance, be an expansion of simple behavioural therapies applied by health visitors under the supervision of clinic-based psychologists. Child psychiatrists could do more sessional work in primary care settings. Community paediatricians could be integrated into primary care, or given a fuller brief to treat and refer. All these schemes, however, would imply an increased input. The next steps should therefore be the comparison of the value and cost-effectiveness of different intervention schemes. They are worth trying. Hyperactivity can be a symptom of several psychiatric problems; but hyperkinetic disorders in childhood lead to long-term disability and challenge child health services to help affected children overcome their handicap.

References

Achenbach, T. M. (1988). Integrating assessment and taxonomy. In *Assessment and Diagnosis in Child Psychopathology*, (eds. Rutter, M., Tuma, A. H., & Lann, I. S.). Guilford Press, New York.

—— & Edelbrock, C. S. (1978). The classification of child psychopathology: A review and analysis of empirical efforts. *Psychol. Bull.*, **85**, 1275–301.

Ackerman, P. T., Elardo, P. T., & Dykman, R. A. (1979). A psychosocial study of hyperactive and learning-disabled boys. *J. Abnorm. Child Psychol.*, **7**, 91–100.

American Psychiatric Association, Committee on Nomenclature and Statistics (1980). *Diagnostic and Statistical Manual of Mental Disorders. (ed. 3)*. American Psychiatric Association, Washington, D.C.

—— (1987). *Diagnostic and Statistical Manual of Mental Disorders, ed 3-R*. American Psychiatric Association, Washington, D.C.

August, G. J., & Stewart, M. A. (1982). Is there a syndrome of pure hyperactivity? *Brit. J. Psychiat.*, **140**, 305–11.

——, ——, & Holme, C. S. (1983). A four-year follow-up of hyperactive boys with and without conduct disorder. *Brit. J. Psychiat.*, **143**, 192–8.

Barkley, R. A. (1982). Guidelines for defining hyperactivity in children. In *Advances in Clinical Child Psychology, Vol. 5*, (eds.. Lahey, B. B. & Kazdin, A. E.). Plenum, New York.

Bee, H. (1967). Parent-child interaction and distractibility in nine-year-old children. *Merril-Palmer Quarterly*, **13**, 175–90.

Berg, C. A., & Steinberg, R. J. (1985). Response to novelty: Continuity versus discontinuity in the developmental course of intelligence. In *Advances in Child Development and Behavior, Vol. 19*, (ed. Reese, H.W.). Academic Press, New York.

Bohman, M., Cloninger, R. C., Sigvardsson, S., & von Knorring, A.-L. (1982). Predisposition to petty criminality in Swedish adoptees: I. Genetic and environmental heterogeneity. *Arch. Gen. Psychiatry*, **29**, 1233–41.

Bosco, J. J., & Robin, S. S. (1980). Hyperkinesis: prevalence and treatment. In *Hyperactive Children: The social ecology of identification and treatment*, (eds. Whalen, C. K. & Henker, B.). Academic Press, New York.

Boyle, M. H., Offord, D. R., Hofmann, H. G., Catlin, G. P., Byles, J. A., Cadman, D. T., Crawford, J. W., Links, P. S., Rae-Grant, N. I., & Szatmari, P. (1987). Ontario Child Health Study: I. Methodology; II. Six-month prevalence of disorder and rates of service utilisation. *Arch. Gen. Psychiat.*, **44**, 826–36.

Buss, A. H., & Plomin, R. (1975). *A Temperament Theory of Personality Development*. Wiley, New York.

Cairns, E., & Cammock, T. (1978). Development of a more reliable version of the Matching Familiar Figures Test. *Devel. Psychol.*, **14**, 555–60.

Campbell, S. B. (1973). Mother-child interaction in reflective, impulsive and hyperactive children. *Devel. Psychol.*, **8**, 341–9.

Campbell, E. S., & Redfering, D. L. (1979). Relationships among environmental and demographic variables and teacher-rated hyperactivity. *J. Abnorm. Child Psychol.*, **1**, 77–81.

Ceci, S. J., & Tishman, J. (1984). Hyperactivity and incidental memory: Evidence for attentional diffusion. *Child Devel.*, **55**, 2192–203.

Childers, A. T. (1935). Hyper-activity in children having behavior disorders. *Amer. J. Orthopsychiat.*, **5**, 227–43.

Cloninger, C. R. (1986). A general theory of personality and its role in the development of anxiety states. *Psychiat. Devel.*, **3**, 167–226.

Conners, C. K. (1969). A teacher rating scale for use in drug studies with children. *Amer. J. Psychiat.*, **126**, 884–8.

Corter, C. M., & Minde, K. M. (1987). Impact of infant prematurity on family systems. In *Advances in Developmental and Behavioral Pediatrics, Vol. 8*, (eds. Wolraich, M. & Routh, D. K.). JAI Press, Greenwich, Conn.

Davidson, L. L., & Taylor, E. (1987). Parental supervision does not explain the lack of association between overactivity and childhood injuries. *Amer. J. Diseases in Children*, **141**, 382.

Deutsch, C. K., Swanson, J. M., Bruell, J. H., Cantwell, D. P., Weinberg, F., & Baren, M. (1982). Overrepresentation of adoptees in children with the Attention Deficit Disorder. *Behav. Genet.*, **12**, 231–8.

Douglas, V. I. (1988). Cognitive deficits in children with attention deficit disorder with hyperactivity. In *Attention Deficit Disorder: Criteria, Cognition, Intervention*. A Book Supplement to the Journal of Child Psychology and Psychiatry, No. 5, (eds. Bloomingdale, L. M. & Sergeant, J.). Pergamon Press, Oxford.

Drillien, C., & Drummond, M. (1983). *Developmental Screening and the Child with Special Needs*. Clinics in Developmental Medicine No. 86. SIMP/Heinemann, London.

Egger, J., Carter, C. M., Graham, P. J., Gumley, D., & Soothill, J. F. (1985). Controlled trial of oligoantigenic treatment in the hyperkinetic syndrome. *Lancet*, **i**, 540–5.

Erlenmeyer-Kimling, L., & Cornblatt, B. (1978). Attentional measures in a study of children at high-risk for schizophrenia. *J. Psychiat. Res.*, **14**, 93–8.

Ferguson, H. B., & Rapoport, J. L. (1983). Nosological issues and biological validation. In *Developmental Neuropsychiatry*. (ed. Rutter, M.). Guilford Press, New York.

Fergusson, D. M., & Horwood, L. J. (1987). The trait and method components of ratings of conduct disorder—Part II. Factors related to the trait component of conduct disorder scores. *J. Child Psychol. Psychiat.*, **28**, 261–72.

Firestone, P., & Martin, J. E. (1979). An analysis of the hyperactive syndrome: A comparison of hyperactive, behavior problem, asthmatic and normal children. *J. Abnorm. Child Psychol.*, **7**, 261–73.

Gillberg, C., Carlstrom, G., & Rasmussen, P. (1983). Hyperkinetic disorders in children with perceptual, motor and attentional deficits. *J. Child Psychol. Psychiat.*, **24**, 233–46.

Glow, R. A. (1981). Cross-validity and normative data on the Conners Parent and Teacher Rating Scales. In *The Psychosocial Aspects of Drug Treatment for Hyperactivity* (eds. Gadow, K. D. & Loney, J.). Westview Press, Boulder, Co.

Goldstein, H. S. (1987). Cognitive development in low attentive, hyperactive and aggressive 6- through 11-year-old children. *J. Amer. Acad. Child Psychiat.*, **26**, 214–18.

Goodman, R., & Stevenson, J. (1989). A twin study of hyperactivity: I. An examination of hyperactivity scores and categories derived from Rutter Teacher and Parent Questionnaires. II. The aetiological role of genes, family relationships, and perinatal adversity. *J. Child Psychol. Psychiat.*, **30**, 671–710.

Goyette, C. H., Conners, C. K., & Ulrich, R. F. (1978). Normative data on revised Conners' parent and teacher rating scales. *J. Abnorm. Child Psychol.*, **6**, 221–36.

Hagen, J. W., & Hale, E. A. (1973). The development of attention in children. In *Minnesota Symposia on Child Psychology, Vol. 7*, (ed. Pick, A. D.). University of Minnesota, Minneapolis.

Hartsough, C. S., & Lambert, N. M. (1982). Some environmental and familial correlates and antecedents of hyperactivity. *Amer. J. Orthopsychiat.*, **52**, 272–87.

Hinshaw, S. P. (1987). On the distinction between Attentional Deficits/ Hyperactivity and Conduct Problems/Aggression in child psychopathology. *Psychol. Bull.*, **101**.

Horowitz, L. M., Wright, J. C., Lowenstein, E., & Parad, H. W. (1981). The prototype as a construct in abnormal psychology: I. A method for deriving prototypes. *J. Abnorm. Psychol.*, **90**, 568–74.

Jenkins, R. L., & Stable, G. (1971). Special characteristics of retarded children rated as severely hyperactive. *Child Psychiat. Human Devel.*, **2**, 26–31.

Kanner, L. (1957). *Child Psychiatry, Third Edition*. Blackwell Scientific Publications, Oxford.

Kaplan, B. J., McNicol, J., Conte, R. A., & Moghadam, H. K. (1989). Dietary replacement in preschool-aged hyperactive boys. *Pediatrics*, **83**, 7–17.

Klorman, R., Brumaghim, J. T., Coons, H. W., Peloquin, L.-J., Strauss, J., Lewine, J. D., Borgstedt, A. D., & Goldstein, M. G. (1988). The contributions of event-related potentials to understanding effects of stimulants on information processing in attention deficit disorder. In *Attention Deficit Disorder: Criteria, Cognition, Intervention*, (eds. Bloomingdale, L. M. & Sergeant, J.). A Book Supplement to the Journal of Child Psychology and Psychiatry, No. 5. Pergamon Press, Oxford.

Koriath, U., Gualtieri, C. T., van Bourgondien, M. E., Quade, D., & Werry, J. S. (1985). Construct validity of clinical diagnosis in pediatric psychiatry: relationship among measures. *J. Amer. Acad. Child Psychiat.*, **24**, 429–36.

Lahey, B. B., Schaughency, E. A., Strauss, C. C., & Frame, C .L. (1984). Are attention deficit disorders with and without hyperactivity similar or dissimilar disorders? *J. Amer. Acad. Child Psychiat.*, **23**, 302–9.

Lambert, N. M., Sandoval, J., & Sassone, E. (1978). Prevalence of hyperactivity in elementary schoolchildren as a function of social system definers. *Amer. J. Orthopsychiat.*, **48**, 446–63.

Langley, J. (1984). Injury control–psychosocial considerations. *J. Child Psychol. Psychiatry*, **25**, 349–56.

Levin, P. M. (1938). Restlessness in children. *Arch. Neurol. Psychiat.*, **39**, 764–70.

Loney, J., Langhorne, J., & Paternite, C. (1978). An empirical basis for subgrouping the hyperkinetic/minimal brain dysfunction syndrome. *J. Abnorm. Psychol.*, **87**, 431–41.

——, & Milich, R. (1982). Hyperactivity, inattention and aggression in clinical practice. *Advances Devel. Behav. Pediat.*, **3**, 113–47.

Luk, S.-L., & Leung, P. W. L. (1989). Conners' teacher's rating scale— a validity study in Hong Kong. *J. Child Psychol. Psychiat.*, **30**, 785–94.

——, Thorley, G., & Taylor, E. (1987). Gross overactivity: A study by direct observation. *J. Psychopath. Behav. Assessment*, **9**(2), 173–82.

McGee, R., Williams, S., Bradshaw, J., Chapel, J. L., Robins, A., & Silva, P. A. (1985). The Rutter scale for completion by teachers: Factor structure and relationships with cognitive abilities and family adversity for a sample of New Zealand children. *J Child Psychol Psychiat.*, **26**, 727–40.

——, ——, & Silva, P. A. (1984). Background characteristics of aggressive, hyperactive and aggressive-hyperactive boys. *J. Amer. Acad. Child Psychiat.*, **23**, 280–4.

Matejccek, Z., Dytrych, Z., & Schuller, V. (1985). Follow-up study of children born to women denied abortion. In *Abortion: Medical Progress and Social Implications*, (eds. Porter, R. & O'Connor, M.). Ciba Foundation Symposium No. 115. Pitman, London.

Nichols, P. L., & Chen, T.-C. (1981). *Minimal Brain Dysfunction: A Prospective Study*. Erlbaum, Hillsdale, N.J.

Pastore, R. A., & Scheirer, C. J. (1974). Signal detection theory: considerations for general application. *Psychol. Bull.*, **81**, 945–58.

Porrino, L. J., Rapoport, J. L., Behar, D., Sceery, W., Ismond, D. R., & Bunney, W. E. (1983). A naturalistic assessment of the motor activity of hyperactive boys. I. Comparison with normal controls. *Arch. Gen. Psychiat.*, **40**, 681–7.

Prendergast, M., Taylor, E., Rapoport, J. L., Bartko, J., Donnelly, M., Zametkin, A., Ahearn, M. B., Dunn, G., & Wieselberg, H. M. (1988). The diagnosis of childhood hyperactivity. *J. Child Psychol. Psychiat.*, **29**, 289–300.

Prichep, L., John, E. R., Ahn, H., & Kaye, H. (1983). Neurometrics: Quantitative evaluation of brain dysfunction in children. In *Developmental Neuropsychiatry*, (ed. Rutter, M.), Guilford Press, New York.

Prior, M., & Sanson, A. (1986). Attention deficit disorder with hyperactivity: A critique. *J. Child Psychol. Psychiat.*, **27**, 307–20.

Quay, H. C. (1979). Classification: Patterns of aggression, anxiety-withdrawal and immaturity. In *Psychopathological Disorders of Childhood*, (2nd ed.). (eds. Quay, H. C. & Werry, J. S.) Wiley, New York.

Quinton, D. (1988). Urbanism and child mental health. Annotation. *J. Child Psychol. Psychiat.*, **29**, 11–20.

——, Rutter, M., & Liddle, C. (1984). Institutional rearing, parenting difficulties and marital support. *Psychol. Med.*, **14**, 107–24.

Rapport, M. D., Tucker, S. B., DuPaul, G. J., Merlo, M., & Stoner, G. (1986). Hyperactivity and frustration: The influence of control over and size of rewards in delaying gratification. *J. Abnorm. Child Psychol.*, **14**, 191–204.

Reeves, J. C., Werry, J. S., Elkind, G. S., & Zametkin, A. (1987). Attention deficit, conduct, oppositional, and anxiety disorders in children: II. Clinical characteristics. *J. Amer. Acad. Child and Adol. Psychiat.*, **26**, 144–55.

Robins, L. (1978). Sturdy childhood predictors of adult antisocial behavior: Replications from longitudinal studies. *Psychol. Med.*, **8**, 611–22.

Routh, D. K. (1980). Developmental and social aspects of hyperactivity. In *Hyperactive Children: The Social Ecology of Identification and Treatment*, (eds. Whalen, C. K. & Henker, B.). Academic Press, New York.

Rutter, M. (1983). Introduction: concepts of brain dysfunction syndromes. In *Developmental Neuropsychiatry*, (ed. Rutter, M.). Guilford Press, New York.

—— (1989*a*). Isle of Wight revisited: Twenty-five years of child psychiatric epidemiology. *J. Amer. Acad. Child Psychiat.*, **28**, 633–53.

—— (1989*b*). Attention Deficit Disorder/Hyperkinetic Syndrome: Conceptual and research issues regarding diagnosis and classification. In *Attention Deficit Disorder: Clinical and Basic Research*, (eds. Sagvolden, T. & Archer, T.). Erlbaum, Hillsdale, NJ.

——, Cox, A., Tupling, C., Berger, M., & Yule, W. (1975). Attainment and adjustment in two geographical areas. I. The prevalence of psychiatric disorder. *Brit. J. Psychiat.*, **126**, 493–509.

——, & Garmezy, N. (1983). Developmental psychopathology. In *Handbook of Child Psychology (Vol. IV: Socialization, personality and social development)*, (ed. Mussen, P. H.). Wiley, New York.

——, & Gould, J. (1985). Classification. In *Child and Adolescent Psychiatry, Modern Approaches*, (eds. Rutter, M. & Hersov, L.). (2nd edition). Blackwell, Oxford.

——, Graham, P., & Yule, W. (1970). *A Neuropsychiatric Study in Childhood*. Spastics International Medical Publications/Heinemann, London.

——, Tizard, J., & Whitmore, K. (Eds.), (1970). *Education, Health and Behaviour*. Longmans Green, London.

——, Tuma, A. H., & Lann, I. S. (Eds.), (1988). *Assessment and Diagnosis in Child Psychopathology*. Guilford Press, New York.

Sandberg, S. T. (1981). The overinclusiveness of the diagnosis of hyperkinetic syndrome. In *Strategic Interventions for Hyperactive Children* (ed. Gittelman, R.). M. E. Sharpe, New York.

—— (1986). Overactivity: Behaviour or syndrome? In *The Overactive Child* (ed. Taylor, E.). The Mac Keith Press/Blackwell, London.

——, Rutter, M., & Taylor, E. (1978). Hyperkinetic disorder in psychiatric clinic attenders. *Devel. Med. Child Neurol.*, **20**, 279–99.

——, Wieselberg, M., & Shaffer, D. (1980). Hyperkinetic and conduct problem children in a primary school population: Some epidemiological considerations. *J. Child Psychol. Psychiat.*, **21**, 293–311.

Satterfield, J., Hoppe, C. M., & Schell, A. M. (1982). A prospective study of delinquency in 100 adolescent boys with attention deficit disorder and 88 normal adolescent boys. *Amer. J. Psychiat.*, **139**, 795–8.

Schachar, R. J. (1986). Hyperkinetic syndrome: historical development of the concept. In *The Overactive Child*, (ed. Taylor, E. A.). Clinics in Developmental Medicine No. 97. The Mac Keith Press/Blackwell, London.

——, Rutter, M., & Smith, A. (1981). The characteristics of situationally and pervasively hyperactive children: implications for syndrome definition. *J. Child Psychol. Psychiat.*, **22**, 375–92.

——, & Taylor, E. (1986). Clinical assessment and management strategies. In *The Overactive Child*, (ed. Taylor, E. A.). Clinics in Developmental Medicine No. 97. The Mac Keith Press/Blackwell, London.

Scott, P. D. (1963). Psychopathy. *Postgrad. Med. J.*, **39**, 1–7 *Et To*.

Sergeant, J. (1981). *Attentional Studies in Hyperactivity*. Rijksuniversiteit te Groningen, Groningen.

—— (1988*a*). From DSM–III attentional deficit disorder to functional defects. In *Attention Deficit Disorder: Criteria, Cognition, Intervention*, (eds. Bloomingdale, L. M. & Sergeant, J.). A Book Supplement to the Journal of Child Psychology and Psychiatry, No. 5. Pergamon Press, Oxford.

—— (1988*b*). RDC for hyperactivity/attention disorder. In *Attention Deficit Disorder: Criteria, Cognition, Intervention*, (eds. Bloomingdale, L. M. & Sergeant, J.). A Book Supplement to the Journal of Child Psychology and Psychiatry, No. 5. Pergamon Press, Oxford.

Shaffer, D. (1980). An approach to the validation of clinical syndromes in childhood. In *The Ecosystem of the "Sick" Child*, (eds. Salzinger, S., Antrobus, J., & Glick, J.). Academic Press, London.

Shapiro, S. K. & Garfinkel, B. D. (1986). The occurrence of behaviour disorders in children: The interdependence of attention deficit disorder and conduct disorder. *J. Amer. Acad. Child Psychiat.*, **25**, 809–19.

Shekim, W. O., Kashani, J., Beck, N., Cantwell, D., Martin, J., Rosenberg, J., & Costello, A. (1985). The prevalence of attention deficit disorders in a rural midwestern community sample of nine-year-old children. *J. Amer. Acad. Child Psychiat.*, **24**, 765–70.

Shen, Y.-C., Wong, Y.-F. & Yang, X.-L. (1985). An epidemiological investigation of minimal brain dysfunction in six elementary schools in Beijing. *J. Child Psychol. Psychiat.*, **26**, 777–88.

Shepherd, M., Oppenheim, B., & Mitchell, S. (1971). *Childhood Behaviour and Mental Health*. University of London Press, London.

142   References

SPSS (1983). *SPSS User's Guide.* McGraw-Hill, New York.

Stewart, M. A., Cummings, C., Singer, S., & DeBlois, C. S. (1981). The overlap between hyperactive and unsocialised aggressive children. *J. Child Psychol. Psychiat.*, **22**, 35–46.

Still, G. F. (1902). The Coulstonian Lectures on some abnormal psychical conditions in children. *Lancet*, **i**, 1008–12, 1077–82, 1163–8.

Swanson, J., Kinsbourne, M. (1976). Stimulant related state-dependent learning in hyperactive children. *Science*, **192**, 1354–6.

Sykes, D. H., Douglas, V. I., & Morgenstern, G. (1973). Sustained attention in hyperactive children. *J. Child Psychol. Psychiat.*, **14**, 213–20.

Szatmari, P., Offord, D. R., & Boyle, M. H. (1989). Ontario Child Health Study: prevalence of attention deficit disorder with hyperactivity. *J. Child Psychol. Psychiat.*, **30**, 219–30.

Taylor, E. A. (1980). Development of attention. In *Scientific Foundations of Developmental Psychiatry*, (ed. Rutter, M.), Heinemann Education, London.

—— (Ed.), (1986). *The Overactive Child.* Clinics in Developmental Medicine No. 97. The Mac Keith Press with Blackwell Scientific, London/ J.B. Lippincott, Philadelphia.

—— (1987). Cultural differences in hyperactivity. *Advances in Devel. Behav. Pediat.*, **8**, 125–50.

—— (1988). Attention deficit and conduct disorder syndromes. In Rutter, M., Tuma, A. H., & Lann, I. S. (eds.), (op. cit).

—— & Sandberg, S. (1984). Hyperactive behavior in English schoolchildren: a questionnaire survey. *J. Abnorm. Child Psychol.*, **12**, 143–56.

——, Schachar, R., Thorley, G., & Wieselberg, M. (1986a). Conduct disorder and hyperactivity: I. Separation of hyperactivity and antisocial conduct in British child psychiatric patients. *Brit. J. Psychiat.*, **149**, 760–7.

——, Everitt, B., Thorley, G., Schachar, R., Rutter, M., & Wieselberg, M. (1986b). Conduct disorder and hyperactivity: II. A cluster analytic approach to the identification of a behavioural syndrome. *Brit. J. Psychiat.*, **149**, 768–77.

——, Schachar, R., Thorley, G., Wieselberg, H. M., Everitt, B., & Rutter, M. (1987). Which boys respond to stimulant medication? A controlled trial of methylphenidate in boys with disruptive behaviour. *Psychol. Med.*, **17**, 121–43.

Thorley, G. (1984). *A fourteen-year follow-up study of severely hyperactive children and psychiatric controls.* Thesis submitted in partial fulfilment of PhD degree, University of London.

Tizard, B., & Hodges, J. (1978). The effect of early institutional rearing on the development of eight-year-old children. *J. Child Psychol. Psychiat.*, **19**, 99–118.

Trites, R. L., & Laprade, K. (1983). Evidence for an independent syndrome of hyperactivity. *J. Child Psychol. Psychiat.*, **24**, 573–86.

Vikan, A. (1985). Psychiatric epidemiology in a sample of 1510 ten-year-old children. I. Prevalence. *J. Child Psychol. Psychiat.*, **26**, 55–75.

Waldrop, M. F., Pedersen, F. A., & Bell, R. Q. (1968). Minor physical anomalies and behavior in preschool children. *Child Devel.*, **39**, 391–400.

Weiss, G. (1983). Long-term outcome: Findings, concepts, and practical implications. In *Developmental Neuropsychiatry*, (ed. Rutter, M.). Guilford Press, New York.

——, Hechtman, L. T. (1986). *Hyperactive Children Grown Up*. Guilford Press, New York.

Werry, J. S., Reeves, J. C., & Elkind, G. S. (1987). Attention deficit, conduct, oppositional and anxiety disorders in children: I. A review of research on differentiating characteristics; II. Clinical characteristics. *J. Amer. Acad. Child Psychiat.*, **26**, 133–55.

Wolff, S. (1985). Nondelinquent disturbances of conduct. In *Child and Adolescent Psychiatry: Modern Approaches*, (eds. Rutter, M. & Hersov, L.). (2nd ed.). Blackwell, Oxford.

Wolkind, S., & Rutter, M. (1985). Separation, loss and family relationships. In *Child and Adolescent Psychiatry: Modern Approaches*, (eds. Rutter, M. & Hersov, L.). (2nd edition). Blackwell, Oxford.

World Health Organization (1978). *International Classification of Diseases* (9th ed.). Author, Geneva.

—— (1988). I.C.D.-10: 1988 Draft of Chapter V. Categories F00–F99. Mental, behavioural and developmental disorders. World Health Organization Division of Mental Health, Geneva.

Yule, W. (1986). Behavioural treatments. In *The Overactive Child*, (ed. Taylor, E. A.). Clinics in Developmental Medicine No. 97. The Mac Keith Press/Blackwell, London.

——, & Taylor, E. A. (1987). Classification. In *Soft Neurological Signs*, (ed. Tupper, D. E.). Grune & Stratton, Orlando.

Zinkin, P. M., & Cox, C. A. (1976). Child health clinics and inverse care laws; Evidence from longitudinal study of 1,878 pre-school children. *Brit. Med. J.*, **2**, 411–13.

Index